Faith Fear Fortune

Copyright © 2024, By Amy C. Shea

Thank you for purchasing an authorized edition of this book and for complying with copyright laws by not reprinting, reproducing, scanning, or distributing any part of it in any form without express permission from the author or The Platform Press. You are directly supporting the author, and the creative process, through the purchase of this book.

This is a **work of fiction**. Names, characters, places, and incidents, although based on the author's true story, have been altered and fictionalized to protect the identities of those involved.

Published by The Platform Press in Toronto, Canada

Connect: instagram.com/faithfearfortune

www.faithfearfortune.com

www.theplatformpress.com

Prologue

Over a single year, I faced abuse, witnessed the suicide of my partner, my best friend Bella died, and I couldn't work because of post-traumatic stress disorder. At the time these events, experienced together, seemed like too much to bear. Having lost it all, staring into the abyss of my empty life, I began to slowly rebuild. Through this story, I share how I got my life back and found joy. It wasn't easy but it's possible and isn't 'possibility' in the face of unfathomable circumstances truly what's needed the most? Fantastic as these events may seem, this book is based on my own lived experience.

The story starts in my hometown of Toronto, Canada, a city for which I have great affection. The names have been changed, purely out of respect for those involved, but I can assure you of the narrative's overall authenticity. As you read this story, I encourage you to have an open mind because not everything in life is as it seems, or is what you've been conditioned to believe.

At the very least, I hope that by reading my story, you're encouraged to consider life's possibilities. I also wish that if you're facing difficulties in your own life, reading these words may offer you some form of comfort. Whatever you're facing, you're not alone and there is a future that's brighter than your past. Follow your faith, get comfortable facing what truly scares you, and you will find the answers you're looking for. You will find the meaning in it all.

Dedicated to my best friend Bella. Thankyou for the inspiration to write this story, and for your neverending love and companionship.

Part 1: Faith

Faith, in the face of all of life's obstacles,
is what will set us on the right path

The sun was lighting up the dusk backdrop against the hazy rows of sailboat masts, clustered together like tiny floating cottages in the cozy harbor neighborhood. Anna had lived here for the better part of her adult life, drawn to the peaceful serenity of the lake because it reminded her of her childhood. In another life, she may have had children or a relationship to tether her to this place, but having neither, she had remained anchored here out of her love for the water.

Anna considered herself somewhat of a nomad who, at the same time, craved the security and warmth of a comfortable life. Her life had been crafted for herself out of sheer hard work and a dash of pragmatism. She hadn't had it easy, by any means, but her parents had instilled the values of hard work and discipline that served her well in her career as a journalist. This diligence and work ethic had granted her an independence for which she was eternally grateful.

She strolled casually along the waterfront with her dog Ella, breathing deeply and inhaling the last vestiges of summer. The ache of the impending fall and winter months, an indelible cloak shrouding the beautiful Monet-like canvas of picnickers and children cradled by the lush greenery. The cicadas hummed a beautifully strung cacophony that mimicked the buzz of the Toronto backdrop of noisy revelers soaking up the last draws of August.

The tension between the abandon of delight and anticipation of winter always left Anna feeling somewhat melancholy at this time of year. It was a feeling that she couldn't quite put her finger on but one that made the stark solitude of the city sting in a painfully aching and uncomfortable way.

    At forty-two, Anna had searched for companionship over the years, longing for someone to share witness to the beauty of life's theater, a travel companion, a confidant, and most importantly a steadfast friend. This hopeful searching had eventually given way to a halfhearted acceptance that she would likely be navigating the waters of life alone. This constantly nagged at her like an unwelcome guest in the background of her mind.

    She considered herself not perfect obviously, but a great potential friend and life partner for the right person. Anna and her dog Ella were, however, a made match and shared an almost psychic connection. Countless nights she had marveled at the perfect creature curled lovingly at the end of her sofa. Ella had been there as a steadfast companion through the highs and lows, the attempted relationships, the job layoffs, the promotions, and the deaths of loved ones. Her quiet companionship was Anna's reprieve from a world that often seemed to be in constant chaos. Ella's love always made sense.

    Her work as a journalist had offered Anna the luxury of travel at a relatively young age. When she was barely nineteen, she had been offered her first real job as a freelance writer for a small publishing house after she had proven her value through an internship. Anna relished the time alone, immersed in the creative flow, and she could barely believe that she was paid to do something she considered her greatest passion. She had also never traveled as a child and was now suddenly jetting off to the United Kingdom and America.

She counted herself fortunate to have found intrinsic joy in a craft that afforded her an adult life and the ability to plant her roots and move out of her parent's home and away from her mother. Anna loved her mother but she kept Anna in a state of constant fear. Her mother also had bipolar disorder but never accepted her diagnosis or sought treatment. Anna desperately wanted a relationship with her mother but after years of disappointment, she had resigned herself to the fact that it would never happen.

Anna's work ethic and drive had been inherited from her father, Janis, whom she looked to as an example from the time she was very young. Her dad was an extremely brilliant man, whose creativity made him an engaging and successful private school teacher. She liked to think that she had inherited a touch of his creativity as well.

As the sun cast red hues over the sky, Anna made her way across the lush greenery of the parkette towards her condo with Ella, their long shadows cast together in flowing unison. Ella trotted straight-legged with a joyful bounce in her step, her shiny silver locks of hair framing smiling white teeth as she stared lovingly back toward Anna.

Indeed, that impulsive day she had taken the Toronto transit to view a litter of miniature schnauzer puppies that summer morning nine years ago had changed her life most wonderfully and unexpectedly. Anna had taken one look at the tiny shivering ball of black and gray fluff cowered in the back of the crate and fallen deeply in love at first sight. Ella had given her an anchoring warmth and purpose, a cherished routine where there had been somewhat of an untethered drift.

Anna passed her key across the electrical panel and entered into the clean, warm but starkly furnished condo. It wouldn't be long before the greenscape painted across the floor-to-ceiling windows in her crisp corner unit would be replaced by the skeletal figures of dark trees stretching towards the blue-gray skies in the foreground of the Toronto skyline.

She had tried to make her home into a cozy sanctuary that would offer solace from the cold winter streets. There was a warm gas fireplace nestled in a nook of her living room beside bookshelves displaying treasures of her travels to other countries. Her travel artifacts were her most prized possessions because they represented the values of what was most dear to her, the richness of travel and discovery. She had purchased a Brazilian hammock for her balcony that overlooked the clusters of sailboats in the harbor, where she would lay with Ella and escape the chaos in the streets below.

Anna loved the home and the life that she had been able to carve out for herself, but she thought again and again of how nice it would be to have a companion to share it all with. Maybe, she thought, it was time to throw caution to the wind and start meeting new people again. Not tonight though, it was late, and the dusk had given way to obscured twilight, casting blue and red hues across the glass skyscrapers and reflecting the last crescent of the setting sun.

She gently carried Ella to bed, lifting the covers and tucking her in gingerly, with a gentle kiss on her warm, hairy forehead. This was a routine to which they had both become warmly accustomed and it was a welcome bookmark before they would begin a new day. She slid under the duvet next to her furry companion and switched off the master lights.

THE NEXT MORNING Anna was up early. She carried Ella to the living room, got her breakfast and brewed herself a fresh cup of espresso. As she mentally prepared herself for the day, she reflected on her fruitless romantic pursuits over the past five years. There was Tom, with whom she'd been in a semi-serious relationship for the better part of two years. Their relationship had erupted into a blaze of flames in the spring.

She had met him through her neighbor Bo, who was part of her sailing club. At the beginning of the relationship, it had seemed fun. He took her to fancy restaurants and events that made her feel simultaneously both an imposter and an admirable socialite. They went out for weekends on his powerboat, drinking wine and watching the sunset. But over the years, it became clear to Anna that she was being used as just another one of his playthings, a distraction to help him cope with his failed marriage of twenty years.

Repeated attempts to discuss building something real like a

home or a family had all inevitably ended in tears and heartache. In the end, she had come to terms with the fact that the relationship would never offer her the equally committed partnership she was longing for. She had ended things rather abruptly, like the ripping off of a weathered bandaid, to avoid compounding the emotional damage. Anna wasn't terribly downtrodden to see him go. The entire relationship, she had felt like she was trying to live up to an ideal she could never sustain.

In the months that followed, she had gone on a few dates. There was the attractive, muscled man named Elias she had met in the gym. There was Matt, who she had been set up with by a former colleague. None of these chance encounters had amounted to anything of substance or even the kindling of a new friendship. Anna had been avoiding the one place she knew there would be an abundance of potential partners: online dating sites.

She had attempted online dating at one point in her life, and this had ended in a string of brief, aching disappointments. Modern city life in Toronto had stripped away opportunities for the organic beauty of chance encounters. If she were to actively pursue a relationship, it seemed that this was her only option. Anna decided she was going to update her profile with recent pictures of her travels and make another valiant attempt. If she could somehow manage to keep her expectations in check, she would be able to mitigate the damages and stay positive through the dating process.

FAITH   FEAR   FORTUNE

AFTER WORK, Anna set to crafting an accurate depiction of herself on Ramble, one of the more popular dating sites. She sat and stared blankly at the screen, her mind vacant. How does someone represent themselves in a few images, capturing a fleeting expression and moment, a few brief sentences that might not even be read? As a journalist, she had been a prolific writer, capturing multidimensional human interest stories, and now she could barely put a few words together to describe herself.

Anna decided to start with the visual aspect of her profile and selected a few pictures of her travel adventures, including one of her following one of her passions: sailing. She filled out a few pieces of generic information on herself—age, location, and checkboxed interests. She knew that men would be unlikely to look at any of this, but she filled it out anyway. Ten meager minutes, and she had put herself out there to be discovered by tens of thousands of men. The thought of being discovered by one of her male colleagues sent a brief pang of anxiety through her. What would they think of her being on Ramble? But she put that fear to rest with the thought that they too were on Ramble, so who were they to judge her?

Anna published her profile and went to pour herself a crisp glass of Chardonnay. She gingerly swiped through a few profiles, wondering how she could get an accurate picture of what any of the men were like, except for the cut of their pectoral muscles in the bathroom mirror. After a few hours of mindlessly ping-ponging back and forth with a few of the men she had matched with, she came across a match named Calvin who seemed rather intriguing.

His profile image looked strangely familiar, and on closer inspection, it seemed to be taken from the waters of her sailing club, with her apartment building in the background. Anna liked Calvin's profile, and was delighted to see that five minutes later they had matched and she had a new message from him in her inbox, "Hello Anna, how is your night going so far? Are you up to anything interesting?"

"Good evening! I'm just relaxing and watching some Netflix after a long day of work. I'm new on this app and just getting my feet wet," Anna replied, deciding to ask about the picture of him at the club, "Looking at your profile photo, it looks familiar! Were you by any chance at the National Sailing Club when this was taken?"

She stared at the glass of wine, condensation dripping gently down the stem, the breath in her lungs hanging in restrained suspension while she watched the reply indicator pulse. The air in the room suddenly felt heavy, and she wondered if she was making a mistake by offering up this information so quickly.

"Yes, that photo was taken at the National Sailing Club. I'm a member there and I took this from out on the mooring ball. Do you know people there?" He replied in minutes.

Anna sat for a few minutes, stunned. How had she come across someone who seemed to share her greatest passion in a matter of minutes, and he also wasn't too hard on the eyes? This, at least, she thought, could be the kindling of a friendship. "Wow, what are the chances? I am a member of the National too! I've been crewing there for ten years or so, and I live in the area. I take it you're also a sailor! Do you know Bo Chatterji? He's a good friend of mine."

"Of course I do! I helped him step his mast in May on launch day; he's a great guy. We grabbed drinks a few weeks ago. We should grab a drink on the patio sometime this weekend if you're up for it," Calvin quickly wrote back.

'Was this really happening right now?' Anna thought. A few hours ago she was contemplating the idea of putting herself out into the dating world again, and in a matter of minutes, she seemed to connect to someone who could be a viable match for her. "Sure, I'd love to grab a drink on the patio, how about Sunday afternoon? Are you free?"

"Sure! Does three work for you? I'm looking forward to meeting you and talking club gossip lol. It should be a really nice day on Sunday too, the August heat is still burning bright. You can also ask all your friends at the club about me to get the inside scoop, I don't mind, hahaha," Calvin wrote.

This seemed safe to Anna, sailing friends in common who were very dear to her. In her experience, sailors were generally good people, especially the ones at her club who needed to get sponsored by a member in good standing and even provide references and a background check to join. "Let's do it! I'll see you there at three on Sunday." Anna set her phone down on the table. She gazed into the speckled, twinkling lights of the city, the soft pulsing orbs of streetlights casting a yellow haze on her periphery, illuminating the dark walkways she would soon lead Ella down. She walked over and gently lifted Ella from her curled haven in the corner of the sofa and cradled her, moving to the entry and placing her into her harness.

"Your mama is going on a date on Sunday, miss Ella! Wish me luck!" Anna cooed to Ella who acquiesced with a gentle blinking of her brown, long-lashed eyes and a flash of her white canines. 'You have something fun to look forward to this weekend, but don't get your hopes set too high. You don't want to add any more disappointments to your roster,' thought Anna as she latched the door and headed outside.

2
The Date

Sunday morning, Anna got up with the hurried momentum of anticipation, excited for her date with Calvin. She walked Ella and then sipped a rich dark cup of her Costa Rican coffee. The azure tones of blue on Lake Ontario sparkled with the morning sun as the bells of the sailboat masts played sweet chords that were carried to her on the breeze. It was going to be a tepid day; the air was already sticky with humidity from the lake, which carried with it the smell of warmed rockweed from the bay. The seagulls shrieked as they swooped in to grab some cereal in the park that had been dropped by a child, also shrieking loudly as her mother pushed her along hurriedly in a stroller.

From her balcony, Anna could see a litany of activity, as throngs of people flocked to the waterfront to drink in the beauty of one of the last Toronto summer weekends. She could see lush rolling greenery dotted by colorful patterned picnic blankets. This fevered activity included cyclists who ripped along the path, being honked at by cars as they skirted the stop signs rebelliously at the intersection beside her home. Longboarders dodged pedestrians and dogs as they glided along the trail gracefully.

Anna's excitement matched the pulse of activity below, and she couldn't wait for her afternoon patio soirée. It had been at least six months since she had gone on a proper date, and she felt a slight tinge of anxiety about how it would go. She had never been that great at small talk during new encounters; her nervousness always seemed to get the better of her.

Anna had dealt with anxiety on the best of days, and dating, well, that was an area that quickened her pulse. In addition to anxiety, she had also been diagnosed with bipolar disorder two years after her first manic episode. After she was finally diagnosed, many events in her life started to make sense, and the past started to reconcile itself finally.

Anna had always been prone to bouts of torrential gushing creativity and seemingly endless energy from the time she was a teen, which inevitably was followed by equally pronounced bouts of deep depression. It had taken her many years and a great number of doctors to reach a diagnosis and then months to find the right cocktail of medications that would moderate her emotional highs and lows while allowing her creativity to flourish. It was an inherited illness she suffered from, like diabetes, and she refused to let it define her.

The stigma around bipolar disorder had initially made her reluctant to accept her diagnosis and seek treatment, but she eventually relented and had nothing but gratitude once she was treated. Her relationships, her concentration, her mood, and ability to remain positive in the face of hardships had all improved dramatically. There were no longer the pronounced highs and lows, but she hadn't lost her creativity. Her quality of life had taken a one-hundred-and-eighty-degree turn for the better.

Anna no longer withdrew from family and friends into the crushing darkness of deep depression regularly. Sure, she could still have melancholy moments, but they didn't feel like they were even in the same realm as the sharp, aching void she was accustomed to. With the medication, Anna was able to pull herself out of dark places and keep moving forward.

After treatment, her confidence was boosted immensely, and she finally felt as though she had a great deal to offer another in terms of emotional stability and care. She had wasted years of her life being undiagnosed, never really understanding why it was so difficult to connect with anyone or why her relationships didn't seem to work out. She felt like she was finally ready to look for a partner with whom she could reciprocate healthily.

Anna set Ella gently down beside the hammock and slipped out of it, balancing her half-full coffee. "Time to get moving, Ella, we have a big day ahead," to which Ella murmured in soft grunts. Anna got into the shower, the warm water running over her face, thinking about how the date with Calvin would go. Anxiety started to tug at her chest, and she decided to focus her attention on getting ready for the day. Thinking about how it might go wouldn't change anything, after all. She was determined to put her best foot forward.

The heat of the day was stifling, and she had chosen to wear a light pair of tan dress shorts and a black top so the perspiration, which would inevitably soak her back, wouldn't show itself so boldly. Anna walked with purpose towards the National Sailing Club, which was a mere block from her home. She cherished having the club so close to her home; it made it easy to be on time for any events or crewing dates she needed to make, and she was extra grateful that it would allow her to be a few minutes early for her date with Calvin. Anna hated being late, especially for dates, because it left her feeling flustered.

A FULL TEN minutes early for their date, Anna climbed the steel grate stairs that would take her up to the patio of the club and decided to pick a seat in the corner that would give them some privacy. It was important that they would get the ability to have an uninterrupted conversation, should that transpire of course. The patio was almost empty for such a beautiful sunny day, and she figured that everyone else must be out sailing.

'Who knows if he's even going to show up?' Anna thought. She gazed across the bay at the masts swaying gently, to the yachts that were drifting peacefully out of the gap towards the western open waters of Lake Ontario. The lake was so large that even on a clear cloudless day, you could barely see to the other side. It had always felt more like an ocean than a lake to Anna. Vast and expansive, it took people from five to eight hours to sail across, depending on the wind and direction of travel.

It was a calm day, and there were only about three knots of wind to carry the vessels forth. Conversely, Anna was feeling voracious ten-knot gusts pressing her heart out of her chest, and she was starting to have second thoughts about the date. She started to have a panicked moment but looked inwards from the lake across the patio as she saw a flurry of male energy bounding towards her, 'This must be Calvin.' she thought.

"Hey, you must be Anna! I hope so; otherwise, this would be really awkward," Calvin piped loudly as he took his seat across from her. Anna smiled as she took note of Calvin's confidence and extroverted mannerisms. 'Very presumptuous of him to sit down without a reply,' she thought.

"Yes, I'm Anna, and you're obviously Calvin! It's so great to meet you. Did you want to grab a beer or something; it's so hot out. It must be over forty degrees with the humidity!" Anna returned nervously. Calvin acquiesced, and they ordered a pair of cold draft beers to quench the heat and take the edge off of their encounter.

"So, you've been a member here at the club for a long time?" Anna piped up nervously before she took a long drawn-out sip of beer.

"Yes, I've been sailing since I was a little kid, and I joined up about six years ago. I know just about everyone at the club because I sail my C&C on race nights and take a lot of the new crew members and social members out to help them discover sailing and get out on the water. I love giving back and helping the club out. Why did you join the club, Anna?" Calvin responded.

"Well, I love the water, and I live right next door. I've been kayaking for about twelve years now, and I always wanted to learn how to sail. I was actually the first kayak member at the club. Something about being on the water brings me a sense of peace in the chaos of the city, and it feels like such an escape," Anna explained.

"Wow, how is it that we have so much in common! I feel the exact same way. My job as a Director of Sales is so stressful, and I get relief when I'm sailing. I come out here by the water and just feel peaceful. We should go out sailing sometime if you're up for it," Calvin prodded Anna for a second date.

"Yes, that sounds fun! Apart from sailing, what else do you like to do?" Anna asked. She couldn't believe she had met someone with whom the conversation seemed to flow so naturally. It was also amazing that they had so much in common, even a shared passion for the water. Could it be that online dating had worked out for her?

Calvin was also from a small town, leaving home with little money and barely the clothes on his back to make his way as a successful Director of Sales in food services. It wasn't just the superficial things that aligned either; he was also a keen health nut who highly valued time with his dog Maggie, and like Anna, he had vulnerably admitted that he suffered from anxiety and had stewed for hours about the date prior to meeting her.

'This definitely has potential. I'd at least love to see where a friendship might go; he seems to share so many interests!' Anna thought after agreeing to a second date with Calvin.

3
The Air Show

Her date with Calvin had lasted for hours, and Anna, feeling elated, walked Ella through Little Norway Park with an extra bounce in her step. Sensing Anna's high spirits, the gray ball of canine joy bounced along in unison ahead of Anna, regularly smiling back at her with gleaming eyes. Tall oak trees bordered an invisible path adjacent to the paved one, and Ella favored this grassy option as it offered her more pee-mail to read.

A self-organized game of baseball was being played by the neighborhood children, who gleefully slid into the golden dust of home plate as the sun set. The next day would be Labour Day, a national holiday, and the standard bedtimes were tossed to the wayside. The cicadas, which seemed to drone nearer and louder from the tall oaks, filled the muggy air, oppressive even at the ripened hours of dusk.

Her phone buzzed in her pocket accompanied by the warm, familiar chime of her favorite childhood video game, Mario Bros. It was a message from Calvin, "It was so great to meet you, and this might be quick, but I'm taking a bunch of friends out to watch the airshow on my boat tomorrow around 10 AM. If you and your friends want to come along, you're more than welcome to. My boat is thirty-three feet, and I can only take maybe three more people with me."

Anna paused, 'This seems to be moving faster than I'm used to, but he did say that he's bringing a bunch of friends and that I can bring my friends along. He also seemed so great, and we did have a lot in common. What have I got to lose,' she thought, breathing deeply.

"It was great to meet you too, I enjoyed our conversation. Sure, why not! It's last minute, but I will see if any of my friends can join up tomorrow. Thank you so much for the invite; it sounds like so much fun. I love watching the air show and usually see it from my balcony, but seeing it from the water would be a real treat," Anna replied.

Anna gazed across the lush greenery of the park to the stately willow that stretched its arms towards couples drinking wine on soft quilted blankets. Scenes like this tended to cast a dull pallor on Anna's heart, but today this was replaced by a hopeful feeling, caressing her heart and warming it. Maybe one day Anna could be enjoying wine in blissful comfort on a blanket with someone special, maybe sooner than she had imagined. She quickly sent messages to her friends Carlie, Hannah, and Mel, to see if they were free to join for a sailing adventure in the morning. To her delight, Carlie and Hannah both responded yes with excitement, fervently inquiring about the captain.

'Today was a wonderful day. I can't wait until tomorrow,' Anna thought as she responded quickly to her friends with her excitement thinly veiled in texts. They would have to wait until tomorrow for the full account of her first date with Calvin, and she couldn't wait to fill them in.

THE NEXT DAY, Hannah arrived at Anna's apartment first. As Anna gave Hannah an amiable hug, she could smell the pungent stale cologne of alcohol on her breath. Hannah had lost her job as a legal assistant because of excessive drinking affecting her work performance a few months prior, and this deep personal disappointment had steered her further off the straight and narrow towards her addiction. Hannah was a good person but she refused to seek or receive help.

Anna knew the friendship was one-sided, and she gave more to Hannah than Hannah could ever offer in return, especially now, but the truth was that Anna felt bad for Hannah and thought she could help her by showing her kindness. Hannah was, after all, funny, smart, and charismatic despite being deeply troubled. Maybe if Anna showed her some love, she would see the good in herself and find a way to change her habits.

'If only Hannah could see all of these positive qualities in herself and stop being so self-destructive. I hate seeing her do this to herself again and again,' Anna thought as she led Hannah to the living room to sit down, suspecting that Hannah was likely already intoxicated.

"So, who is he?x Where did you meet him?" Hannah chirped as she pulled a three-liter box of wine from her canvas sack and popped it on the table. "We are going to have so much fun today, I dragged this box of wine all of the way from the subway station and my arms are burning so we have to drink it now so it weighs less to carry over to the club," she continued as she popped off the red wax cap and poured herself a warm, generous glass of Pinot Grigio.

Anna swigged the last sip of coffee and sighed, 'How could she possibly want to be drinking wine at nine in the morning? When did she start this morning? It was a mistake extending the invitation to her. She's going to make an ass of herself,' she thought. Anna couldn't focus her attention on Hannah today though, she needed to steel herself for her date and prepare mentally.

"So, I actually met him through Ramble, but he's a member of the club and he knows my neighbor Bo. You met Bo at a party last summer; he is so kind and really smart. Bo said he's a really good guy, and I trust Bo to be a decent judge of character," Anna explained to Hannah as she poured herself the tiniest glass of wine to hold in pained solidarity with her friend. She didn't want to drink before her date.

"Seriously, that's so awesome, and he has a boat, and he's taking us out today. We're going to have such an amazing time girl, and if he's a jerk at least we have a nice day out on a boat. Maybe he has some friends for me. Shane and I are still seeing each other, but I've also been seeing my neighbor Greg," Hannah twittered in between sips.

"Ugh, don't you have enough going on with two men? Maybe

you should focus a bit more," Anna offered.

Before Hannah could respond, Anna's phone chimed its familiar tone, and she buzzed the caller into her apartment foyer, "That must be Carlie," she said. 'Saved by the bell,' she thought.

A few minutes later, as Hannah continued her expeditious inebriation, Carlie arrived bringing with her a usual quiet calm. Carlie couldn't be more opposite from Hannah, but she tolerated Hannah for the sake of her friendship with Anna. Carlie was a health researcher, and a brilliant, kind woman.

"Hey, guys, sorry I'm a bit late. The streetcar was taking forever today because they are doing construction on Queens Quay," Carlie offered with an apology. Carlie had been Anna's friend since they were nineteen, and Anna considered her to be a haven of support and genuine caring. They were the rare kind of friends that could go without speaking for months and come back together as though not a minute had passed.

"Don't worry we still have lots of time to get over to the club, but are you guys ok if we leave soon? I don't want to be late," Anna asked.

"Yeah, I can take a traveler. Carlie, do you want a glass of wine before we go?" Hannah quipped as she chugged back the rest of her wine like it was a glass of water.

"No, I'm good. I just want to get a glass of juice, and then we can go. I brought some cheeses and fruits that we can take over and share," Carlie fended off the offer of wine in better form than Anna had done. Carlie never showed up empty-handed for an event, and it was just one of the things Anna loved about her.

The trio left the condo with portable coolers and snacks, and minutes later, they were at the club. They searched for Calvin's sailboat, 'Necropolis,' amongst the mismatched fleet nestled cozily along the wall. "Ahoy!" Calvin yelled loudly from down the line of boats. "Great to see you, Anna, and that your friends made it as well! We should get going so we can beat the police blockade in the harbor. We can all meet and do introductions on the boat," he continued.

"Awesome, thank you again for having us out. We are so excited for the day!" Anna returned as the three of them stepped onboard clumsily to find the few spare patches of deck amongst Calvin's friends, who all seemed to be in great spirits.

The hull of the craft was weathered and cracked but beautiful. Anna thought all boats were beautiful simply because of the freedom they afforded their owners. It was clear Calvin loved his boat and took good care of it.

"Let's set sail, crew!" Calvin called as two of his friends, Rick and Val, cast off the lines and stepped briskly back onto the boat.

Stealthy gray crafts etched hazy smoke patterns into the crystal-clear sky, preceding the offset, bellowing drum of the engines that were so loud they seemed to rock the small boat back and forth and press them against the deck. The group of eight, including Captain Calvin, were crowded together as they strained their necks to see the impressive machines rip in close proximity past them.

Hours passed as they watched the dance unfold above them, too loud to do anything but sip drinks and recline against the curved deck, exchanging excited smiles and a few elevated words of awe. Introductions were limited to first names and occupations. In addition to Anna, Carlie, Hannah, and Calvin, there were four of his friends. Val was a stuntwoman, Rick was a police officer, Lauren was still in the military, and Polly was an actress, all of them in their late twenties, attractive and scantily clad.

Anna felt out of place amongst Calvin's group of friends but reasoned that Calvin was even older than her and looked even more out of place. Anna's friends were more her breed of people, and they were wearing actual clothing, like her. She didn't understand why he was interested in her if these were the types of friends he kept but tried to push that feeling down.

As the show ended and Necropolis tacked a zigzagging path back to National Harbour Club, Calvin pressed Anna for a third date, "I feel bad that we didn't get to talk much today, but I'm glad I got a chance to see your sailing skills. Can I take you out for dinner on Thursday, and you can meet my dog, Maggie? I like your friends, but I like you even more. You are a pretty amazing person, Anna, in case no one ever told you. I can see myself falling for you fast."

Anna felt her stomach flutter, and warmth rose to fill her cheeks, "I'd love to grab dinner with you, and I'd love to meet Maggie. I'm sure my dog Ella would love to meet her too. Thanks again for bringing us out today; that was nice of you. Your friends seemed really cool."

"Yeah, most of my friends I met by helping new crew members come out. Ok, it's settled then, I will be at your house at about seven on Thursday, and there's a little Italian place on King Street that we can go to; it's one of my favorites. I am really glad that you and your friends had a good time today," Calvin said, sealing the date.

As the group exchanged niceties and goodbyes, Calvin and Anna smiled warmly at each other and embraced. 'I can feel myself falling for this man; he's just so amazing, and I can't believe that he seems to feel the same way in return,' Anna thought to herself, 'Could he be the partner I've been hoping for after so long? I hope so. It seems like we already have a great friendship brewing. Maybe my lonely days are over.'

Anna felt light as she glided back to her apartment with Hannah and Carlie. Hannah's typically annoying drunken tittering didn't even register as mild annoyance with Anna now; she was on cloud nine, and nothing could bring her down.

4
The Move

Thursday evening came quickly, and the dinner date was a continuation of the deep connection Anna could feel building between her and Calvin. Over a bottle of Chardonnay and generous plates of homemade gnocchi, he expressed a familiar brand of pained loneliness to her that was accentuated by the prolonged pandemic. The pandemic and social distancing had been enduring worldwide for over a year.

Lockdown protocols had taken their toll on Anna and many of her friends, isolating them from friends and family and straining relationships. The standard loneliness of being solo in the big city was amplified many times over during the Coronavirus pandemic. Anna had cried frequently, missing her friends and family, completely cut off from the outside world. She was so grateful for Ella's companionship during this painful time.

Like Anna, Calvin had expressed the yearning for a relationship and had also lamented over the challenges he had faced in his recent dating pursuits that were made worse by the lockdowns. He had detailed his last relationship of five months with a woman named

Amanda, saying it had ended after she hurled kitchen items at him in a fit of rage. He described her as "crazy" and told Anna that she wouldn't cease her efforts to contact him. He told Anna she would call him all the time, making up excuses like she missed his dog Maggie or worryingly, that she was going to hurt herself.

'Poor Calvin,' Anna thought. She was no stranger to painful and abusive relationships, even having had to obtain a restraining order against one of her unrelenting past partners when she was in her twenties. Anna felt deep empathy for Calvin's situation and experiences, not dissimilar to her own. As the date neared the end, Calvin leaned in to kiss Anna and she was glad to let it happen. It seemed right to Anna, so right. Anna let Calvin drive her home that night, elated, feeling as though she had finally met her match.

IN THE WEEKS that followed, Calvin and Anna, as well as their dogs, were seemingly inseparable. Calvin had seemingly limitless time to devote to Anna and expressed that he had never been more happy in his adult life, and had never experienced the intense emotions he was feeling for Anna. Initially apprehensive about the speed the relationship was accelerating, Anna had relented against her fears, and she was allowing herself to fall for Calvin. She desperately wanted and needed it to be true.

AFTER ABOUT A MONTH, one evening following a particularly emotionally intense conversation, Calvin suggested they try living together to see how it goes. "When you're our age, you just know what you want, and I want you, Anna. I know and feel this deeply like nothing I've ever known before," Calvin had expressed his intentions to Anna during dinner. "I have a fantastic job, making six figures, and we both love sailing, and it just makes sense for us to be together," he continued to make his case.

Anna thought for a few days about his suggestion and decided that if he kept his place as a fallback option, in the event it didn't work out, then she didn't have anything to lose. She also didn't want to keep wasting time in her life. A trial of living together would let her know truly what kind of person Calvin was to be with. So Anna agreed with the caveat that he would keep his place and only bring what was necessary.

IT WAS A CRISP Fall day when Calvin arrived with Maggie in the front seat of his blue Honda Civic packed to the brim. The leaves were piled in decaying clumps along the curb, and the sun was low in the sky as they pulled up. Anna greeted Calvin nervously when she saw the sheer amount of stuff he had brought with him. "Hey, if I'd known you were bringing so much over I would have organized a little better. I don't know if I can even fit all of that in my condo. This is a trial, right?" Anna stuttered.

FAITH   FEAR   FORTUNE

All of a sudden, the walls of the buildings along the narrow street seemed to be getting closer, caving in towards Anna slowly, making her dizzy. Calvin offered reassuringly, "Don't worry this isn't as much as it looks like; it's mostly clothing, and you'll have to start calling it our place now darling. We're in this life together now right? I'm committed to doing this with you; you mean everything to me."

This was all starting to send Anna into a quiet panic, but she tried to hide her anxiety with a wide forced smile, "Yes, that's right. Let me take Maggie upstairs and then come back down to help you bring some stuff up." Anna had run from relationships in the past when she was scared, and now she was in her forties, it was time to stop running away once and for all.

Anna tried to force aside the fear that was welling up inside of her stomach. "I'm so excited that we're going to be doing this together Calvin, I am. Sorry, I'm just a bit tired, I didn't mean anything by that," she offered with an apologetic tone. The two of them proceeded to carry up Calvin's things, making eight trips each. Anna's small condo was piled with items, but it was late, so they left the organizing for later.

"I love you, Anna," Calvin said for the first time after they were done, and out walking the two dogs together. "I love you too," Anna instinctually responded but then felt a deep pang of a strange feeling she couldn't quite wrap her mind around.

The next day, Anna started moving around some of her stuff to make more space for Calvin's things. She wasn't particularly attached to her things and started a pile for donation. If anything, this would be a chance to declutter her space a bit. She had emptied the spare closet the previous week, in anticipation of his arrival, but it was clear she hadn't cleared enough space. 'This is a small price to pay for having someone to share my life with,' she thought as she worked away.

BY THE END of the day, the condo was looking well-organized, and most of Calvin's things had found a place. To celebrate, Calvin had surprised her with champagne and takeout sushi from her favorite restaurant. 'He is quite thoughtful, I think I might have mentioned that restaurant once, and he remembered," Anna thought to herself, 'I need to stop being so afraid all the time.'

After their third glass of champagne, Calvin suggested that they were both feeling way too stressed out from work and it would be a great idea for them to go away to spend more time together to start living together off on the right foot. Anna was used to taking at least two all-inclusive vacations a year and was overdue for one. "Where do you like to travel to, Anna? I'd like to experience one of your favorite places with you. We can go somewhere cheap and cheerful," Calvin said.

"Well, I do love Cuba. I've been there more times than I can count, and it's so relaxing and affordable. The beaches in Cayo Coco are stunning. The people there are also so friendly, which always makes them feel like home to me," she replied.

"I've been to Cuba and I loved it there too! I only saw Havana, but I'd love to see another part of the island. Let's look now and see if we can find something on sale, at the last minute," Calvin pressed. Anna always loved the idea of travel, so she fired up her laptop and found an all-inclusive package with a flight leaving in just two weeks.

"I can get the time off work, so let's just book it," Calvin said, "I will send you a money transfer and pay you back tomorrow for my half." This seemed to make sense, so Anna pulled out her credit card and processed the trip for both of them, knowing that getting the time off work would be a breeze for her. Anna was way overdue for a vacation, and she couldn't say no to a tropical beach destination.

5

The Proposal

The next weekend, Anna and Calvin set out with their two dogs for an overnight sail to Snug Harbour on Calvin's sailboat, the Necropolis. This would be their first full weekend away together in the small sailboat with the two dogs, and one of the last sailing weekends of the short Toronto Summer. The days were still warm, and the high sun gleamed off the white sails, reflecting on the lake, glittering like diamonds against the black depths of Lake Ontario. They cut a path through the water, making a steady five knots, and were into the safety of Snug Harbour by dusk, just as the rains came.

Overnight a storm rolled in, and the tiny boat rocked back and forth against the wooden dock, banging loudly and frightening Ella and Maggie. The two dogs whimpered together under the covers of the main berth, while Anna tried to sleep. Calvin had already passed out hours earlier. It was supposed to be a fun, relaxing weekend but the winds and rains made it more a harrowing journey than a peaceful escape. Anna spent the entire time worrying about Ella and Maggie.

AFTER A WEEKEND of rough sailing against strong gusts,

Calvin admitted that his boat probably wasn't the most comfortable for long-haul trips with the two dogs. "You know Anna, you really ought to come to check out Jenny's boat with me. She just put it on the market, and it's less than ten years old, and she's been good to her boat. There's no harm in looking right?"

It was indeed Anna's dream to own a sailboat ever since she was a young child, and this was amplified when she moved to the tiny harbor neighborhood twenty years ago. After giving it some consideration, she agreed that there was indeed no harm in looking. "Sure I'd love to check it out. You know that's a lifelong dream I've had since I was five. We grew up on the harbor in Oakville, and I think it's in my blood," she acquiesced.

Back at the National Sailing Club, Anna and Calvin meandered down the main dock towards Jenny's yacht, which was indeed a stately and beautiful thirty-four-foot craft, gleaming in the sunlight of the late afternoon with her sails nestled into new navy sail covers. Anna stepped onto the yacht and felt the energy of freedom and the clarity of her lifelong dream closer than ever before. As she made her way into the cabin of the vessel, she fell in love with the warm beveled woodwork, everything purposefully built and intricately placed. In an instant, she saw a bright future of sailing trips across the lake, evenings sipping wine, watching the sunset.

"I love her, I do, but I don't know how to sail a boat this big, and I've never owned a boat. The price is right, and I could afford it, but it's a lot of boat for me," Anna exclaimed to Calvin in disappointment. As much as she wanted her own boat, it was also a terrifying proposition, to have to take care of it on her own. It would be a huge responsibility that she didn't want to bear the weight of on her own.

"If you want her, I will make darn sure you know everything about sailing her, mark my words. If you want to do this, we will do it together," Calvin reassured Anna in a loving tone, "I know Jenny has another offer on the table right now, but if you pay cash with no conditions, I'm pretty sure she would take your offer to avoid any headaches," he continued.

Anna thought about it for a few hours, sitting pensively on the picnic table beside the clubhouse, weighing the pros and cons in her head. 'This is everything I've always wanted and it seems fated that it's happening now. This is the sailboat of my dreams, and I'm finding it just as I've met the man of my dreams. Everything feels right,' she thought. Anna finally decided that it was a passion and a dream that she didn't want to let slip away out of fear. She rang Jenny on her cell, "Hi Jenny it's Anna, will you take a draft cheque? I can drop it off at the club tomorrow."

TWO WEEKS LATER, the air was frosty and thick with moisture as Anna set out to take Ella and Maggie for a morning walk. It was too cold for anything but a thick down jacket, and at seven in the morning, it was still dark outside. The faint haze of yellow sunrise was beginning to force its way up into the darkness of night, and a dewy mist could be seen rising from the baseball diamond. Anna wasn't a fan of the shorter Fall days, but it didn't bother her now because she was leaving for Cuba with Calvin the next morning. In no time she would be on the beach, sipping a cold beer with the sand between her toes, the sun shining high and inviting.

Her friend Cole had graciously offered to watch the two dogs while they were away, saying that it would be a great trial for his family to teach his children about the responsibilities of dog ownership. Cole's family had been hounding him to adopt a dog for years. Anna looked lovingly at Ella, who seemed to adore her new friend, Maggie, trotting alongside her in bouncing unison, and felt comfort knowing that they would be together while she and Calvin were away traveling.

THE NEXT MORNING came early, and Anna's alarm clock rang at the ripe hour of four, letting her know that it was time for her and Calvin to rise and make their way to the airport. Their flight was leaving at seven forty-five, which meant they would have a full, beautiful day in the white sandy Cuban paradise. The trip itself was straight south for three hours. Anna had taken the trip herself many times to flee the cold urban concrete in favor of the azure ocean and salty breeze. She loved traveling to Cuba on her own; it was safe, and she always met such wonderful people there.

As Anna and Calvin boarded the plane with the throngs of sunny destination travelers, she noticed that he was acting rather nervously. "Are you okay Calvin? Are you nervous about flying?" Anna inquired.

"Yes, this is just our first trip together, and I'm excited. I can't wait to sit my butt down in a beach chair and grab a rum. I want to get going. I'm certainly not nervous about flying in an airplane," he laughed off Anna's questions with a casual air.

Satisfied with his explanation, Anna boarded and then took her window seat, reclining against the thick plane window blurred by condensation. She quickly drifted to sleep, awakening only as the plane started to make its final descent into the arid air over the island of Cuba. Anna couldn't believe that they were already there; it always amazed her that in such a short time she could enter such a different world, such a warmer, friendlier place.

As she gazed across the white cresting waves of the Atlantic, which came sharper and more vividly into focus, the plane banked hard for the runway. They slammed into the cracked, sun-weathered tarmac with a harsh bounce, and the packed plane, filled with joyful passengers, clapped and cheered. This surprised Anna and she exclaimed, "Did they think we weren't going to make it or something? You don't hear this when you land in Paris or London."

"Oh, don't worry about it, we're here," Calvin retorted dismissively. Anna shrugged off his reaction, just happy to be on sunnier shores.

Varadero was a bustling beach haven for tourists from all over the world seeking refuge from their mundane, routine lives. Canadians, in particular, tended to flock to the sunny beaches of Cuba in droves every year during the Fall and Winter months, searching for calm waters and sweet escape. Cuba was an attractive destination partly because of the low cost but mainly because of the stunning quality of the white sandy beaches.

BY DAY THREE, Anna had drifted into a familiar blissful state of relaxation, cold beer in hand, turquoise waters glimmering endlessly in front of her. Calvin had seemed to be riding the same wave of relaxation, but suddenly he got very agitated and restless. "Let's go for a walk," Calvin said.

Anna, feeling content reading her novel, reluctantly agreed, "OK, I wouldn't mind exploring that section of the beach to the south where the hurricane hit three years ago, the area that's abandoned. As long as you give me a chance to re-coat myself in sunblock first."

They navigated around the dilapidated remnants of antiquated lifeguard towers, white paint weathered off by the incessant pounding of salty hands. They explored abandoned resorts carefully in their bare feet, picking their way around the hurricane-battered wooden beach bars. As they walked back and neared their resort, Anna noticed that there was a man with an iPad who seemed to be filming them, "What the heck is that guy doing, why is he filming us, that's so rude, I'm going to tell him to stop," she said annoyed.

"No, no, don't do that please. Just one moment," Calvin got down on one knee and sloppily produced a solitaire diamond ring from his swimsuit pocket.

"What! Really?" Anna exclaimed in surprise, "I had no idea that you were planning to do this. How did you keep it a secret?"

Calvin continued, "Will you marry me? I love you."

Anna looked back at the man with the iPad who was now joined by a larger group of white-haired retirees who were all smiling widely from ear to ear, a captive audience. "Ok, yes?" She said nervously as Calvin pushed the ring onto her finger, jumped up, and wrapped himself around her forcefully amid cheers from the group of strangers on the beach who had produced a bottle of champagne and small plastic cups.

"How long were you planning this?" Anna asked incredulously.

"Oh, just since last week. I was listening when you said you wanted to get married someday Anna, and I couldn't picture spending my life with anyone else," Calvin returned.

"Hello dear I'm Beth, congratulations!" a smiling, portly gray-haired woman offered as she placed a plastic cup of champagne into Anna's hand, "Getting married was the best decision of my life, Brian and I are celebrating forty-five fabulous years on this holiday, he's that bloke over there with the iPad. I'm so happy for you two."

"Thank you so much, and nice to meet you, Beth!" Anna replied, not quite being able to put her finger on the subtle feeling of dread that seemed to creep its way up her body. Something didn't feel right, but she didn't have much time to think about it as the group of onlookers, now blasting Cuban salsa on a portable Bluetooth speaker, grabbed her hand and started twirling her around in celebration. 'Isn't this everything I had ever wanted?' she thought as Calvin smiled warmly back at her.

6

The Monster

When they arrived back home three days later, the stark contrast of the frigid Toronto air and wind was a shock to Anna. Somehow it had seemed much warmer at home before their voyage down to Cuba even though only a week had passed. They had just two more weekends before haul-out, when the yachts would be lifted carefully out of the water by cranes that seemed to render the multi-ton vessels weightless. There was a very limited amount of time to enjoy Saoirse, the name she had christened the stately yacht, meaning 'freedom' in Gaelic.

Calvin and Anna donned thick down-filled coats, thermal gloves, and windproof pants to take Saoirse out for her first maiden sail together. They were wearing the matching navy ball caps that Anna had ordered, with the logo she designed for Saoirse embroidered across the front. They had arrived while they were away on vacation. On the dock, Anna marveled at the yacht and couldn't believe that it was hers. She had worked so hard for so many years to make a life for herself, and this seemed to be the culmination of all of that effort.

"Ready to go?" Calvin said curtly. Anna started to unwrap the bowline from the bollard and walked it back to the center of the boat, maintaining tension to ensure it wouldn't swing away from the docked position.

"What the heck are you doing, Anna? That isn't how you're supposed to take off the lines," Calvin snapped angrily at her, "I've got it, it's fine. Just go sit at the helm and back her out real slow."

Anna was taken aback by Calvin's tone and cursing and went to sit at the helm as instructed. Calvin gestured at Anna to start the slow reverse out of the slip. Anna wasn't used to the throttle, and she gave it a bit too much power. Calvin threw the lines on the boat, leapt on deck, and bodychecked her forcefully away from the helm while quickly grabbing the wheel and throttle. "Hey, give me a break, it's my first time backing out," Anna said with annoyance.

"You know everything, Anna, you're an expert, right? If you want to crash your boat, it's your problem," Calvin said angrily at her, his face contorted into a sneer. Anna hadn't heard him swear or raise his voice at her before, or seen this expression of rage on his face, and it sent a pang of heavy terror into her chest.

As they pulled out of the safety of the break wall, Calvin glared at Anna from the helm with an air of superiority, "It's not personal. I've been doing this all my life, Anna, and I know a lot more than you do, and you're going to have to learn to take direction better."

"How is it not personal! You just pushed me off the helm of my boat. You said you would teach me. I intend to learn how to manage

Saoirse myself, and I won't be able to learn if you don't let me make mistakes. I didn't hurt anything," Anna reasoned. She couldn't believe she was having to plead her case to drive her boat.

"It's our boat, Anna, stop being a bitch. I sold Necropolis so I could do this with you. I put everything on the line for you. I gave up my lease and sold my furniture and now I have nothing, Anna. We are in this together, remember," Calvin screamed.

Anna sat in horror as she stared back at Calvin, panicked. 'Who the heck is this person?', she thought, 'He gave up his lease, we didn't talk about that, we didn't agree to any of this. And it's my boat!' Everything started to feel suffocating. "I paid cash for the boat and it's in my name!" Anna screamed back, "I will not be treated like this, no one talks to me this way, certainly not my fiancé, Calvin!"

Anna went to sit on the bow and cried silently as the wind blew her tears back to the stern; she hoped they would pelt Calvin so he would understand the sting that his words had caused her. As they pulled into the harbor, Anna felt like she had the wind knocked right out of her. She caressed Ella and nestled her face into the soft comforting crook of her neck. She moved in slow motion, exhausted by the confrontation and dumbfounded at the change in Calvin's personality. Anna wasn't a confrontational person by any means, but he had drawn out a need to defend herself fiercely. She needed a break from Calvin and was pleased when he suggested that she take the dogs and meet him back at the condo.

FAITH   FEAR   FORTUNE

Anna felt instantly lighter when she started walking home with Ella and Maggie; she tried to rationalize what had happened. 'Did I provoke him?', she thought to herself. By the time she had reached the park with the dogs, she was feeling like herself again, composed and calm.

AN HOUR LATER, Calvin entered the condo in a flurry of activity, arms full, carrying a twelve-pack of vodka coolers. He swiftly plopped himself down on the sofa and cracked a can open, taking a long gulp of the lime-flavored cooler.

Anna was busy organizing laundry in the bedroom, and as she finished and entered the living room, she noticed four empty cans already sitting on the coffee table, lined up in a row. "Why don't we order some nice dinner? We could get a pizza or something, what do you feel like having?" Anna offered in peace.

"I'm not hungry, I just want to watch some shows and relax on the couch now," Calvin said curtly.

"Ok. I'm sorry about what happened on the boat earlier. I don't know how it came to that, but I'm sorry for my part in it," Anna offered. Calvin briefly glanced up from his phone and said nothing in response.

Anna made herself a grilled cheese and tomato sandwich, went to lie down in the bedroom, and texted her friend Hannah, "Hannah are you there? I need some advice. I had a huge fight with Calvin today, and I don't know what happened; he was different, and I barely recognized him. Even his face changed, and it scared me." She fell asleep to a true crime documentary on Netflix waiting for Hannah's reply.

THE NEXT MORNING, Anna woke to the mellow drone of music on her television, the intermission of the streaming service left to run too long. Her phone was still in her hand, and she saw the notification from Hannah, "Don't worry about it girlfriend. Just sleep on it." 'Thanks Dr. Phil,' Anna thought to herself as she rolled over. Calvin wasn't beside her. She went into the kitchen to find a note from Calvin, "Walked dogs. I went to the office late tonight. Sales event at the Hilton."

Anna felt the cold tone wafting off the blunt message and wondered what had made things turn so bad so quickly. She texted Calvin, "Hey hun, hope everything is alright, thanks for walking the dogs this morning I really appreciate it." In a matter of minutes, Calvin curtly replied, "Yup. Fine. See you later."

OVER THE NEXT few weeks, Calvin's demeanor towards Anna didn't improve, and their relationship started to crack apart at the seams, despite Anna's repeated attempts to mend the gap that was opening between them. She had tried to calmly open the lines of communication many times, always offering a heartfelt apology first, and each time it had quickly devolved. Anna didn't know how to talk to Calvin without sending him into a fury, and it was getting worse by the day.

ONE PARTICULAR TUESDAY, after Anna had finished a particularly demanding day at work and was spending time relaxing on the sofa with her feet up, Calvin confronted her aggressively, "Anna! What is it with you? Why are you just sitting around, you've been sitting all day, why don't you get up and move or something? This isn't good for you."

"Please, I'm tired, just let me relax for a bit, and then I will go to the gym around eight. I don't feel like moving right now. It's been a long day of conference calls and stressful meetings, and I just need to decompress for half an hour," Anna said in an exhausted barely audible tone.

"Why are you taking that tone again with me? I never do anything right, I never do anything right do I? I'm a piece of shit for caring, right Anna?" Calvin screamed as he began pounding his chest with his fists like a primate, getting inches from Anna's face.

Anna's anxiety boiled over at this display, and her flight response kicked into high gear. She sprinted toward the bedroom, slamming the door behind her in terror and confusion, and balled herself up on the bed. Calvin cannoned into the room screaming in her face, "I never do anything right, do I!" Anna sprung up and tried to slip by him for the relative safety of the bathroom as she smelt the familiar pungent, stale air of alcohol assaulting her senses. Calvin grabbed her arms and pinned her against the wall causing pain to shoot down her arms.

"Let me go!" Anna screamed in terror, twisting her body down out of his grasp and away as she ran to the bathroom. She slammed the door and pressed the entirety of her body weight against the back of the door as she sank to the floor and sobbed uncontrollably into her hands, wishing she could hold Ella and wondering if she was alright. 'Why is this happening?', she thought. She needed to get away. The situation wasn't right and it was clear there was nothing she could do to put it right; she needed help fast.

IN THE MORNING she went to see her doctor, a trusted confidant of over ten years. "Dr. Nellie, I don't know what to do. I've let my partner move in too fast, I made a mistake and now we're engaged and it's become abusive. I don't even know this person and I'm terrified. He went from being loving and caring to a complete monster, and I think he has a problem with alcohol. He drinks constantly and I don't know how to handle it by myself," Anna broke down in heaving sobs, as Dr. Nellie came over to embrace her.

"Thank you for telling me all of this. You need to know that it takes a lot of courage to talk about these things and I'm proud of you for coming to me. He is emotionally abusing you, that's pretty clear to me. That's bad enough, but has he hit you Anna? Has it become physically abusive?" Dr. Nellie asked measuredly.

"Well, he restrained me and he hurt my arms, and I was terrified, he didn't hit me. He trapped me in the bathroom for hours and I was crying on the floor in my own home. I'm really scared of the situation, it's really bad," Anna said in a quiet voice.

"That's physical abuse, Anna. He doesn't have to hit you for it to be abuse. It's also coercive controlling behavior. I need you to get yourself out of that situation as soon as possible, and I will help you in any way that I can. I have an obligation as a care professional to report any instances of abuse to the authorities. It is for your safety, emotionally and physically," Dr. Nellie continued.

"Can you hold off on contacting the authorities, Dr. Nellie, please? Can I get him to come see you? He needs treatment for his alcoholism. He isn't a bad person, but it's clear that he has a problem. I feel guilty abandoning him; he is sick and he needs to get help," pleaded Anna.

"I don't like this, and I don't like seeing you this way. You're a strong woman, Anna, and you need to be strong now. I respect you, so if this is how you want to proceed, I will honor it. I want to see him in my office this week, though, or I am calling the police. You don't have to tell him why, but if you can get him to call and make an appointment, I'll hold off until I've seen him," Dr. Nellie agreed.

DAYS LATER, Anna was able to get Calvin in to talk to her doctor under the guise of speaking to her to get something for his anxiety, and this gave her hope that things might get better. After he returned to the condo, she tried to speak with him about what happened. "How did the appointment with Dr. Nellie go? I'm here to support you if you want to talk about it, Calvin. I'm here for you," Anna had offered.

"It was really helpful, Anna. Thanks for getting me in. We talked about strategies to help me deal with you and your bipolar, and I think I understand you now. I do. Thanks for connecting me with her; she's a great doctor, and she clearly cares about you," he said with an air of pompous superiority. 'He's never going to take responsibility for himself. This is a nightmare,' Anna thought.

"I got you the appointment because I wanted you to talk about your excessive drinking, Calvin. I am very worried about you, and I genuinely think you need to get help. You're a wonderful person behind all the alcohol, but you have been scaring me lately and—"

Anna was cut off by Calvin yelling, "I have to drink because of you, Anna! I need something to help me deal with your bullshit! I'm the one with all the problems right, Anna? You never deal with your own issues! You need me here to take care of you; you're useless and ridiculous so stop deflecting! That's hilarious, I have a drinking problem of all things!" Calvin paced around the condo like a trapped animal, a vocal stream of insults spewing from him as he slapped himself on the forehead, "I never do anything right. I never do anything right. I never do anything right."

Anna couldn't believe what she was seeing and knew that she had made a terrible error in thinking she could help him. If she'd spent years trying to help Hannah to no avail, how could she help Calvin? Staring at him in horror, she grabbed Ella, ran for the safety of the bathroom, and slammed the door. She assumed her familiar position of sobbing on the bathroom floor, but at least this time she had Ella and was grateful for her company.

OVER THE NEXT week, things didn't improve, and Anna knew that she needed to find a path out. There was no salvaging the relationship. She spent hours trying to formulate a plan to get Calvin to leave that wouldn't cause him to boil over in a rage, one that she could no longer bear the presence of. In the end, Anna decided that she would try to bribe Calvin to leave her. He wasn't going to send her like a battered wife, crying to the authorities; he wasn't going to break her. She thought she could handle matters on her own.

He had firmly entrenched himself in her life in a matter of weeks, and she felt trapped. Every time she would bring up the idea of them going their separate ways, he would scream, "I gave up my life for you; this is what you wanted." Any warmth he had once displayed to her had vanished into thin air. Calvin's face had even changed, and she could barely recognize him. He seemed to wear a permanently contorted sneer, corners of his mouth downturned in perpetual disgust at her and life.

The practical household agreements had gone out the window. Calvin had claimed to make a generous six-figure salary before they moved in together, but Anna had yet to see a penny of contribution from him to either her mortgage or utilities. Anna also had not seen a penny he had promised from the sale of his boat to help maintain Saoirse. She didn't want to admit to herself that she had been manipulated because she thought it meant that she was vulnerable and naive. Anna had been manipulated though, and she needed to find a way to cut her losses and get out. She desperately needed to save herself quickly.

ONE EVENING, when Calvin seemed to be in a relatively good mood and didn't seem to be drinking, Anna approached him and pleaded, "We both know that this isn't working out for either of us, and I know what you gave up, so I would be happy to help you out by paying for an apartment for you for six months and getting you furnishings. We need to go our separate ways for each of our lives to be positive."

"I'm not abandoning you, Anna. You need me. I gave up everything for you. I won't let you do this to me!" Calvin firmly replied.

"I don't need this; you're making me miserable! We're both miserable!" Anna pleaded more firmly, trying to rationalize her way out of the situation, "If it's about money, I said I would pay you; it's ok."

"You don't get it, Anna; I'm not going anywhere. I gave up everything for you and we're getting married!" Calvin raised his voice loudly.

"I can't marry you; I'm miserable and I'm scared of you!" Anna screamed, taking off the ring and throwing it down on the ground in frustration and utter defeat.

"You don't wanna marry me huh?" Calvin said in a low snarling vicious tone, his eyes black like the bubbles of tar pits as he sprung up from the sofa and shook Anna vigorously, dragging her towards the balcony with him, his face deeply contorted. Anna felt a strong hand on her chest that knocked the wind out of her, causing her to slip on the kitchen floor and out of Calvin's grip. She fell hard on two palmed hands as she looked up to see Calvin tip himself backward over the railing of her seventeenth-floor balcony.

ANNA SHRIEKED a guttural sound at the top of her lungs, lunging toward the railing as she stared down in horror at his lifeless body on the mezzanine, blood beginning to seep from the corners of his body. Yelling at the top of her lungs, she looked down at the expressionless onlookers in the parkette, "Calvin, no!" She ran sock-footed into the hallway, screaming and ramming the elevator buttons hard with the butts of her fists. 'Where is the mezzanine, what floor is the mezzanine, nine, eight, ten?' Anna couldn't think or breathe as the elevator seemed to swallow her whole.

Careening through the halls of the first floor, Anna threw herself through the glass doors. She bolted through the lobby toward the security desk, screaming, "I need an ambulance, please, there has been an accident, my partner fell off my balcony, he's hurt, please hurry. He's on the mezzanine." As she was screaming, two women entered the lobby, and one of them grabbed her arm. The pretty brunette said, "We saw what happened, we're here with you, why don't we go upstairs and wait for the ambulance there?"

THE KIND STRANGER led Anna back up to her apartment. To Anna, it seemed like she was in some sort of a nightmare. There were flashes of lucidity that were not strung together in any way that made sense. She was back in her condo, on the floor feeling the thick pile of the gray carpet through her fingers as the police and paramedics came in, and five uniformed officers surrounded her. 'Why are the police here, what is going on? What happened?', thought Anna.

"Why are they here? Where is Calvin? Did you take him to the hospital? Which hospital is he in?" she pleaded with the police officers. An officer with the bluest eyes she had ever seen came to kneel with her on the floor, and he said words that didn't register for Anna, "The paramedic tried everything, that man standing over there worked on him for half an hour, and there was nothing he could do to bring him back. I am so sorry."

Anna couldn't absorb what she was hearing. He had only fallen six floors to the mezzanine; there was no way this could be true. "No, there has to be a mistake; he didn't fall that far from my balcony, just a few floors, no that's not possible," Anna screamed as she lunged for the balcony and then tried to pry open the sliding glass door. One of the officers caught her around the waist and pulled her back.

The next thing Anna recalled was being inside an ambulance. "Where are we going? Are we going to the hospital to see Calvin?" she said softly; she couldn't move her arms. "You're in shock, just stay still. Just know that there was nothing you could have done. My best friend committed suicide during the pandemic, and we had no idea he was going to do it. I understand what you're feeling right now. It's not your fault," the young paramedic tried to reassure her. Anna blacked out.

7
The Faith

Anna awoke in the emergency ward, surrounded by people, most of them strangers. Her friend Cole was the first person she recognized immediately. Among the group of strangers she didn't recognize, there were two uniformed police officers, two nurses, a plainclothes officer, and a medical student.

"Shit Anna, I'm so glad you're alright! I thought it was you who was hurt, and they wouldn't tell me anything on the phone," Cole said with relief, "Your brother is at your condo with the dogs; he and your dad will look after them for a while while you're recovering. They went through your phone and they contacted a bunch of people, including me. Carlie is on her way too. I'm so sorry Anna, I can't imagine what you're going through," he continued.

"What happened? Who are all of these people? I don't understand what's happening here, Cole. I can't remember anything," Anna said in confusion.

"Anna, Calvin's dead. He jumped off your balcony, in front of you, the asshole. These people are the police detectives; they want to talk to you. I told them that you're out of it and won't be able to tell them too much, but they still want to talk to you," Cole explained.

"What? No one is dead, Cole; what are you talking about? This is a nightmare. This can't be happening, where is Calvin? I'm leaving to go figure all of this out." Anna droned off, pushed her way off the stretcher, and started running towards the sliding glass hospital doors. She fell over and blacked out just as she reached the glass doors.

HOURS LATER, Anna woke up for the second time in the back of a minivan, and groggily asked, "Where am I? Whose van is this? Carlie, is that you?"

Carlie grabbed her and gave her a firm embrace. "You're finally awake, Anna," Carlie said warmly, "Don't worry; we're going to take care of you, Anna; everything is going to be alright now. They wanted to send you home right away, so we're taking you to stay near your dad's house for a little while," she continued.

"I don't know what happened, Carlie; I'm really scared. I can't remember anything that happened," Anna exclaimed through tears of frustration. Anna had never been more frightened in her life. It was as though she was in a lucid nightmare, and she couldn't remember the events that had brought her there. She fell asleep on Carlie's lap in the back of the van to the view of highway headlamps fading out.

ANNA WOKE UP in a neatly made bed. The walls were newly painted a rich sage green, and there was a small triangular white desk area built into the corner where the walls met with a tiny window. She could see a gentle covering of snow on the windowsill, and the lack of shadow on the wall outside gave her the impression that it was midday. Out of the corner of her eye, Anna saw a small orange button on the wall beside the bed sparsely labeled with the word, 'Call'.

Anna tentatively pressed the button, and a few minutes later, a round-face, kind-looking nurse entered the room. "You're awake, you poor dear. Are you hungry at all?" the woman asked, "I'm Nurse Sutherland, and if you need anything at all, just let me know, ok. You've been through such an ordeal these past few days," she continued.

"Where am I?" Anna inquired.

"Oh, you're at the St.Catharines General Hospital. Your friends and family wanted to make sure that you were watched and cared for properly. The Toronto hospital wanted to send you home, so they drove you down to us, and we are very happy that they did. You had quite the experience, and they thought you might want to talk about it with a professional. They are worried about you; we all are, Anna," Nurse Sutherland offered, "There isn't anything to worry about now. We are going to help you here, and you can stay as long as needed."

Anna couldn't remember what had happened, but Nurse Sutherland seemed so kind and this put Anna at ease. She felt, somehow, that she was lucky to be there. 'If only I could remember what happened to me. Something Cole said about Calvin is bothering me but I can't remember what it was,' she thought to herself. Anna was tired; she didn't know why she was so tired. She closed her eyes and drifted back to sleep in a matter of moments.

THE NEXT DAY Anna got a phone call, and the nurse brought her cell phone into the room for her to receive it. "Anna! They called me two days ago, and they told me what happened. I am so sorry, hun; I've been trying to reach you; I don't know what to say. I got an email from someone saying you'd be back at work next week; absolutely not! With what you've been through, I've already talked to HR, and you can take at least a month away. Whatever you need, we are here for you," the kind voice of Nola, her former boss, was on the other end of the line.

"I don't remember sending you an email," Anna replied, "I'm really sorry; I don't know what's happening. I can't remember what happened, can you tell me?" she continued.

"I'm so sorry about Calvin's passing. You must be distraught. You can take as much time away as you need. I've already talked to your current boss Andy about it too, and he agrees," Nola expressed her condolences.

"Thank you, Nola; I miss you," Anna managed, confused. She couldn't remember what had happened to Calvin, but she was just then remembering that Cole had also mentioned he was dead. She reasoned that the cause was pretty bad, if Nola had called her directly on her cell phone. She also couldn't remember where she was.

"Excuse me, please nurse, where am I?" she asked the nurse again.

"Oh, you're in the St. Catharines General, Dear," the nurse offered patiently.

Anna couldn't shake the sense of déjà vu that she'd heard that a few times before. She stared up at the speckled white paneled ceiling tiles with their concave pock marks sprayed in a constellation. The warm room was styled like a hotel room but still had a faint chemical hospital smell wafting from the hallways. Anna felt that she had been here before.

STILL IN A STATE of confusion, she was greeted by a woman who entered the room and introduced herself as Dr. Morris. "I'm a clinical psychiatrist, Anna, and I wanted to talk to you about what happened a few days ago. We think it might help you to talk about your experiences. You've seen a lot, and the nurses tell me you're having a difficult time recalling what happened," she explained.

"I can't remember what happened at all, Doctor. People keep telling me that Calvin is dead, but I don't remember what happened to him. Can you please tell me what's happening to me and what happened to Calvin!" Anna said in a frustrated tone, "I'm sorry; I don't mean to be rude, I'm just so scared."

"I'm sorry too, Anna. You went into a state of shock a few days ago, and memory loss is common. It's your body's way of protecting itself. It's your mind's way of protecting itself from overload. You don't need to apologize to me for anything. You're in an acute state of trauma response; go easy on yourself," Dr. Morris said sympathetically.

"Can you please tell me what happened?" Anna pleaded again with the doctor, hoping for answers.

"Well, I spoke to your family, and they told me. Calvin committed suicide in front of you. He jumped off your balcony while you were watching, and it may have just been too much for your mind to process. You were taken to Toronto General Hospital and discharged after a few hours. Your family and friends were rightly concerned and decided to try to find some proper care close to them. I'm glad they brought you here, Anna. I also spoke to your boss, and we agreed that you're going to need some prolonged time away to process everything," Dr. Morris explained further, "It's normal not to remember everything after a traumatic event. It will be ok. You have a lot of support."

ANNA SPENT FIVE more days in the St.Catharines General Hospital, talking to Dr. Morris and the nursing staff, trying to remember everything. Only pieces of the event could be retrieved, flashes of terrifying imagery that Anna pushed away as they sent her into a physical panic. She was ultimately discharged by Dr. Morris into the care of her father, Janis.

"Anna, we were all really worried about you, darling, but we're glad you're alright. I'm personally glad that terrible man didn't harm you. I wouldn't wish that on anyone, but I'm glad he can't hurt you anymore, dear," her dad greeted her.

Anna wanted to be alright because she didn't want to worry anyone, especially her father. She nodded and replied only with, "I love you, Dad. Let's go home if that's ok with you."

WHEN ANNA ARRIVED with her father back at her Toronto apartment, they made their way through the lobby and up the elevator to her floor. Anna felt ill as they made their way down the hall towards the apartment, and she didn't know why. She could remember running sock-footed here, not very long ago, and the thought terrorized her.

There were flashes of bright police and ambulance lights in her periphery that made her dizzy. As they walked down the hallway, paint fumes filled her nostrils, and she could see swaths of yellow tape wrapped across her door, sealing it shut. "Do not cross? What is this, Dad? Why have the police sealed off my home?" she asked, confused.

"I'm not sure what's going on here, hun. Let's go talk to the concierge. I'm sure he will be able to tell us what's going on," Janis replied. They made their way down to the lobby and saw Vishnu, the security guard, sitting at the austere mahogany desk. He was framed by a castle of packages yet to be picked up by residents.

Anna asked, "Excuse me, Vishnu, do you know what's happening with my unit? There is a bunch of police tape across my door sealing it shut."

"Hi Anna, it's nice to see you back. The detectives are still completing their forensic investigation, and I have unfortunately been instructed to not let anyone enter the apartment. I'm very sorry this is happening to you, Anna. I know how difficult this must be for you right now," Vishnu said compassionately, "Lots of different people have been in and out of your apartment for two weeks, and the white cube van parked in the loading dock is theirs as well. No one is supposed to park there either, but there they sit for god knows how much longer!"

"Thank you, Vishnu. I'm sorry this is happening. Thank you for putting up with all of this. I will find out what's going on," Anna said apologetically. "Dad, what are we going to do?"

"It's ok, Anna. Let's take the dogs to the boat, and we'll figure things out from there. We can at least sleep there for a few nights. It's safe, and we can order some groceries for us and food for the dogs, not that you probably feel like eating right now," Janis assured her and set the plan for the next few days. The dogs, Ella and Maggie, were starting to look hungry and restless pacing around each other in the lobby.

Anna felt safe on the boat, so she agreed that this sounded like a good plan. The boat was relatively small, but it would have to do. While the police were carrying out their investigation of Calvin's death, a few days turned into a few weeks. The four of them, Anna,

Janis, Ella, and Maggie, were nestled into the boat like sardines. Anna had awoken multiple times a night, sweating and in terror, and had called the emergency distress line. They had proven to be her lifeline, a voice in the darkness of night.

WHEN IT WAS time to return to the condo, Janis and Anna had driven his small green hatchback to the police station to retrieve the keys to Anna's apartment. Police Constable Park greeted Anna at the front desk of the fifty-two division with a small white envelope with her keys, as well as a larger letter-sized one. "Hi Anna, I'm really sorry to hear about your recent hardships and for any inconvenience that our investigation may have caused you. Please consider the matter closed."

"Thank you, sir. I know that you were only doing what you needed to do. It's alright," Anna responded.

"The coroner will be contacting you in a few days with the results of the autopsy report," Officer Park explained as he handed the envelope with Anna's keys back to her and started opening the larger one, "There is another matter that Calvin's immediate family wanted me to address with you. They have provided a list of items that they want returned that were owned by Calvin. Now, the dog is on the list, but you're under no obligation to return her, it's really up to you."

Anna stared at the stark, white, single sheet of printer paper containing a single columned list of items. 'Was this really what

Calvin had been reduced to? A sheet of paper with a list of material items? I can't believe that they are not even going to help me sort through Calvin's things. What kind of people are they?' she thought coldly to herself. She remembered the stories that Calvin had told her about growing up, the abuse, the arguments. He had once explained how he had been struck so many times his back bled, and now these people wanted his things from him, and they wanted his dog.

Constable Park, taking note of Anna's pained expression, offered, "You don't have to do anything with this right now. Just take it home and think about it. If I understand correctly, you were not married, so you're under no obligation to cover his debts either. You should consider yourself very lucky. Call me if you need anything," he said as he extended her his card.

ANNA AND JANIS drove back to the condo solemnly with the keys and the list of items. Anna turned the lock and opened her condo, cracking the seal of the police tape. She ripped off a few pieces of police tape that took the paint with them and opened the door all the way. Taking a deep breath, Anna turned to look at her father, breath held stationary in her lungs as she walked in.

Moving further into the condo down the hallway, Anna noticed a discarded latex glove in the corner of her living room lingering like an unwanted, disfigured guest. Her father ran over to the balcony window and pulled down the blinds to obscure the view before Anna could look, "You won't have to deal with any of this today, honey," he asserted.

Everywhere she looked, Anna saw reminders of Calvin. Her stomach turned itself in knots every few minutes, and it was all becoming overwhelming. "Dad, I don't know what to do about all of this stuff," Anna sighed in agony as her insides burned.

"We'll just get rid of it fast, we'll fill his Civic with everything we can for the family and get rid of the rest so it's not torturing you," Janis offered sympathetically, "I will be right here by your side, honey. You're not alone. You can do this; there is no rush. Why don't you call a friend to help out?"

"I'm not putting anyone else through this hell. I can get through it," Anna said in resignation. Her father Janis had recently had a hip replacement, and knowing he was in constant pain, Anna wouldn't allow him to help. In truth, she didn't know how she would get through clearing out all of Calvin's things; he was a pack rat and had already filled her space. Anna had to dig deeper than she ever had before to start making labored trips down to the donation bin.

WHEN IT WAS all over, it had taken her a besieged two full days to clear out all of his things from both the condo and the storage locker. There was still the matter of all of his stuff left to go through on the boat. A distraught Anna finally relented and called for help, and her friend Cole came to help her empty the boat. This had taken them another full day. Anna was emotionally and physically spent.

Unable to sleep in her condo because of the flashbacks and terrible feelings, Anna and her father were still reluctantly staying on Saoirse with Ella and Maggie. She had tried desperately, multiple times, to take Saoirse out of the slip by herself and had ended up breaking down in sobbing resignation on the dock. "I don't know what to do, Dad," Anna cried, "I never expected to have such a large boat on my own. I can't do this right now, Dad."

"Maybe it's too much for you right now, Anna. Do you think perhaps it's a good idea to get rid of it and start again fresh with something new next year or the year after? Maybe you should think about making new memories. It's ok if you're not ready for this right now because you didn't expect to be doing it on your own," Janis offered.

"It's my dream, but you're right. This is too much for me right now when I can barely take care of the dogs from lack of sleep. I don't know how I'm going to handle winterization, cleaning, painting, and maintenance. You're right, Dad. You're always right," Anna agreed with her father and contacted a broker who was a trusted friend of the sailing club's Commodore to list the boat. Three days later, the boat was on the market. Anna cried hard five days later when it sold. It felt like the death of a dream that had barely begun to bud.

ANNA RETURNED to work after the month was up. She had

been so busy taking care of the logistical issues surrounding Calvin's death she didn't have time to think about what had happened. She still couldn't remember the events of the day, despite the many explanations from friends and family. Her only glimpse of the event still came as flashes in nightmares that she couldn't escape from, in the dark recesses of her consciousness. The nightmares didn't make sense. The images were horrifying and jumbled together in a dizzying array, waking her abruptly in terror and keeping her awake for the rest of the night.

Anna developed a mortal fear of falling asleep and being subjected to the horror picture show in her mind. Her work performance started to suffer from three or four hours of broken sleep a night, and emotionally, she was barely keeping it together most days. During a performance review, she broke down in tears and felt that she was truly losing it. In desperation, Anna went to see her doctor again, describing her symptoms. Immediately her doctor diagnosed her as suffering from post-traumatic stress disorder. "How can I have trauma if I can't remember what happened?" Anna had questioned.

"The mind works in mysterious ways to protect itself, Anna," Dr. Nellie had explained. Anna had heard this before. "Perhaps, Anna, after everything you have been through, you should consider taking some more time off work to focus on your mental health and well-being. The added stress may be making it difficult for you to make a full recovery. There is no shame in needing to take off; you've been through hell," Dr. Nellie continued.

Anna didn't want to take time off. Her career was everything she had worked so hard for, but at the same time, she could feel herself slipping, and she didn't want to jeopardize the job that she loved. She stared at the anatomy posters of reproductive organs on the wall and hesitantly concurred, "Yes, you're probably right. I'm not functioning like I should be at work. I can't remember things from the previous meetings without writing them down, and to be honest, I don't even really recognize myself right now. I'd probably be doing them a favor by taking time off." Anna felt sick. Her life, everything she had worked so hard for, felt like it was slipping through her fingers like sand.

"Great, you can print off the paperwork for the insurance company and bring it to me. You don't have to do it today; take a few days, and we can go through it together during a follow-up appointment. See the receptionist on the way out, and she'll book you in," Dr. Nellie replied warmly, "I think you're making a really wise decision, Anna. I really do."

Anna's father had returned to St. Catharines a few days prior, so Anna had re-entered her apartment alone after her doctor's appointment. Except, she wasn't completely alone and was grateful for Ella and Maggie's sweet little faces and excited tails-wagging expressions of blissful greetings. After walking the two bundles of love, she set out to print off the paperwork that Dr. Nellie had requested, to get it over with, and called the office to set up a follow-up appointment. Anna was exhausted and curled up on the sofa with Ella in one arm and Maggie in the other.

AROUND EIGHT, Anna had woken up from her snooze on the sofa and taken the dogs out again. As they walked down the Martin Goodman Trail on their way back, Maggie turned to attack Ella for no apparent reason, lunging for her neck, sending Ella whimpering and cowering behind Anna. "No Maggie! Bad Maggie!" Anna scolded her and cradled Ella to her. It was clear that Maggie was missing Calvin and acting out, but Ella was her baby and there was no way she would allow any harm to come to her.

She rang her father in a panic, "Dad, I don't know what to do. Maggie attacked Ella and hurt her. She cut her ear, and I am distraught. I can't have Ella get hurt; she is the love of my life. I'm so exhausted and I don't know what to do about any of this. Your wisdom would be appreciated right now, Dad," she exclaimed to her father.

"Honey, I'm sorry this happened to Ella. I know what Ella means to you, and she was there first and needs to be protected. She's getting older and can't handle the stress. Why don't you consider giving Maggie back to the family; they wanted her anyway. Then it can go back to being just you and Ella again. Do you want a constant reminder of all the pain that you've been through? Doesn't she remind you of Calvin?" her dad reasoned.

"You're right, Dad. It's the right thing to do for everyone right now. Ella is getting older and she deserves to have peace, and I need to protect her," Anna agreed and hung up the phone. She quickly grabbed her laptop and sent an email and let Calvin's family know that they could come and get Maggie any time they wanted. They arrived hours later, and Anna cried as they took Maggie away. She had grown attached to her, and seeing her leave was heart-wrenching.

THE MONTHS PASSED, and Anna fell into a forced routine, devoid of work, devoid of a social life. Anna fell into a deep depression. Most days she didn't see the purpose of getting out of bed in the morning, other than to care for her sweet dear girl. Ella, the tiny ball of gray perfection curled on the end of her bed, was carrying her through the toughest moments of her life. Anna lay in bed, marveling at the sweet, empathic creature that somehow had seemed to be there for her through it all even when people had betrayed her. Anna would force herself to move every morning because Ella needed her.

Anna would get up each day, give Ella her medication, feed her, cuddle her, walk her, and sing to her softly. Anna cherished Ella with all her heart, and was now more grateful than ever to have her quiet companionship. Anna didn't have work, a partner, her sailing passion, her memory, much sleep, or her positive outlook, but she still had a purpose in taking care of Ella. Anna's friends checked in on her every few days, but feeling a mere shadow of her former self, Anna stayed mostly isolated. On her birthday, Cole and Hannah had dragged her out to her sailing club for a pub night where she met a wonderful couple, Mellie and Carl. They had heard about Calvin, and kindly offered to take her out sailing again to get her out on the water.

Anna was reluctant to accept their invitation, but she eventually relented and regained a sense of hope that she had all but forgotten by being out on the water sailing with them. She considered her new friends to be a rare breed of angels who had offered her a lifeline in a time of great darkness. Slowly, Anna began to claw herself out of the lightless pit of depression, and over the months, she started accepting extended hands of support from friends both old and new. She had Ella, and she had support. Anna counted herself grateful for both and would steer her mind towards these blessings, away from the negative thoughts that plagued her. It was bad, but she was still alive; she would never give up on life no matter what happened. Ella needed her.

IT WAS A SUNNY picturesque Sunday, and Anna had accepted an offer from Carl and Mellie to go out on the water. Ella was getting older, and the cloudless sun was too hot on her thick coat of hair. Anna's dog walkers and friends Brent and Dee had agreed to watch her for the day. The wind was light and gentle, and they made two or three knots as they glided along the water, stopping at noon to enjoy cheese and charcuterie before heading back to the club. Anna was starting to remember herself again; she was starting to remember the beauty in life through the fog of her post-traumatic stress disorder, and depression was lifting. She truly felt that she was going to be ok.

Anna carried her bag home, filled with uneaten snacks, and unneeded emergency rain gear. She entered her apartment feeling, for the first time in almost a year, a lightness of being. She dropped half of her bags gingerly in the foyer and wandered down the air-conditioned hallway to the main area of her condo, dumping the bag of unused clothing on her bed to sort through and put away. Anna grabbed her phone and texted Brent, "Hi Brent, I hope Ella's been a good girl. What time do you think you'll be able to bring her back home tomorrow? I miss her so much." She proceeded to start organizing and putting away her clothing and made herself a green tea in the kitchen.

IT WAS HALF an hour later, and Brent still hadn't responded when the phone suddenly rang. Anna picked it up immediately, seeing that it was from Brent. "Anna, are you sitting down?" Brent said softly, "Ella has passed; she went to sleep beside me on the sofa, and it was peaceful, Anna. Dee and I didn't even know until we went to let them out in the backyard." Anna couldn't respond, feeling as though she had been run over by a tractor-trailer she fell onto the floor in a heap and began wailing harder than she ever had before.

'How could she just be gone, Ella no, Ella I love you so much, how can you be gone', the thoughts burned like a charred branding iron on her mind, and her body hung limply against the wall. After an hour she dragged herself to the bedroom and cried until morning, agony forcing the fluid from her body but unable to drink or eat.

THE NEXT MORNING, completely numb, Anna sat on her sofa and stared out the window after she and Brent had taken Ella's body to the vet. A monarch butterfly landed on her balcony, perched just in her periphery, slowly extending and retracting its saturated orange wings. It was the only color in Anna's view, the only color in her life. She closed her eyes to fight back the tears, assaulted by the beauty that Ella would never witness again. The weekend prior, they had walked together to Trillium Park and were surrounded by monarchs. 'Is Ella here with me?' Anna thought to herself, 'Please Ella show me the way, I feel so lost without you.' She pleaded silently.

Her friend Cole had arrived hours earlier and sat in quiet solidarity. There was nothing he could say or do, but his mere presence gave Anna comfort. Carl had arrived the night before for a while to offer solace as she sobbed uncontrollably. There was nothing anyone could do to ease her pain, other than be there in her orbit to help her carry the sorrow.

"Cole, what am I going to do? The walls are caving in on me, I can't stay here. I'm going to die," Anna said through watery tears as she stared longingly at the butterfly, hoping it was Ella reincarnated somehow.

"I don't know, Anna. We all loved Ella, and I know what she means to you. Maybe you should get out of here for a while. Do you think that sitting here is doing you any good after everything that has happened in this place? You need a break, Anna," Cole offered.

"Yes, yes, I need to go far away. I need to leave; I need to save myself. I can't handle this loss. I have nothing left, Cole! Nothing! She was the one that kept me going. How am I supposed to go on without my dear love Ella? My poor sweet girl!" Anna started wailing in sorrow again, her body shuddering with deep sobs.

"Yes, I think that's a good idea, Anna. Get away for a bit. Go somewhere to get some space. Is there anywhere that feels safe for you?" Cole asked.

"Nowhere is safe. I feel like she is still here, Cole! When I was coming back here, I looked up and saw graffiti on the wall that said 'You are not alone.' Who spray-paints that on a wall? I've never seen butterflies up here; why am I seeing them now?" Anna reasoned. "Ella, tell me what I'm supposed to do now. Where do I go from here?" she asked into the ether.

As if guided by some mystic unseen force, Anna grabbed her laptop and Googled, 'What is the significance of monarch butterflies?' The results sent a chill up Anna's spine: 'The monarch symbolizes a loved one coming to offer comfort after their passing and is a common symbol in Guatemalan and Mexican mythology meaning hope and peace. Mexicans, in particular, believe a monarch is a loved one coming back to offer comfort.'

Could it be that Ella was coming to comfort her from beyond the physical realm? Their last day out had been filled with monarchs; was she sending a symbol Anna would recognize and believe? Moved by the symbolism of what was unfolding, she decided Ella was giving her a message and confirming she should leave, and that Anna should save herself.

Pulling up a new window, she Googled, 'Last minute tours from Toronto.' The first result was an adventure package from Toronto to Guatemala, Belize, and Mexico. Anna felt as though she was being led to Guatemala by Ella, and she impulsively booked the trip within minutes. "My life isn't over. It's not over; I'm going to Guatemala," she said to Cole.

FAITH   FEAR   FORTUNE

Cerro de la Cruz, Guatemala

Part 2: Fear

Rebuilding our confidence by facing our fears will enable us to receive life's fortunes

A [Anna]s she had turned the lock on her apartment door before leaving, she felt as though she was fleeing the scene of the wreckage and the carnage that her life had become. Her job was like the twisted metal of a car interwoven with the burned-out body of her love life. As she was leaving, she could feel Ella somehow accompanying her to strange unknown lands; she was truly with Anna in spirit, all around her. It was a strange feeling, and almost as though the air around held some invisible weight to it, instinctively she knew it to be true. Ella was there.

Anna arrived promptly before dawn at the Toronto Pearson International Airport with only a carry-on backpack, of which she had used every available square inch of space. Her travel anxiety had always limited her to what she could carry in the small pack, and she had spent hours organizing and trimming items from her list of must-haves. Every time she began packing, she remembered her earlier days of lost luggage and ruined trips, her bathing suits and toothbrush lost in transit. Minimalist packing was always hard, and it was a welcome distraction for her, to train her mind to the task at hand.

Although traveling this way out of self-imposed necessity, Anna did love the simplicity of traveling with only a carry-on pack. This minimalist way of travel always underscored the unnecessary burden of material things in her life. Encumbered by pain and loss, Anna relished the opportunity to travel as light as possible, shedding her dense physical life like a snake sloughing off the dead skin on the side of the road. Her things were beautiful, yes, but they didn't make her who she was. Anna was the same woman, and had the same value, in a pair of simple black sandals and neutral clothing as she was in a pair of fancy heels and a dress.

THE AIRPORT was bustling at five in the morning. People pushed carts packed with luggage that was arranged like a precarious game of Tetris, as their children screamed and tugged at their parents' coat-tails. All seemed distraught from having been denied a decent night of sleep. Anna was glad to have her tiny pack, and also that she had checked in the night before. She proceeded directly to the security line, and in her travel light state, she was through to the gate in only twenty minutes, take-out coffee in hand. Getting a morning coffee was a ritual for her, and it created a feeling of warmth and coziness no matter where she was.

As the plane lifted gently into the skies, through the layers of stratus clouds, Anna's gaze lingered across the cotton landscape of the sky. She felt only contentment and a lightness of being. Her mind was clear, and she went to sleep in the crook of the seat and the cool, gentle curve of the airplane window, comforted by the droning white noise of the loud jet engines. Cocooned here, she dreamt of happy adventures in foreign lands; positive images filled her consciousness for the first time in many months.

FAITH   FEAR   FORTUNE

ANNA WAS AWOKEN suddenly as the lights were turned on in the cabin accompanied by two loud dings from the cabin crew's announcement system, "The captain has indicated that we have begun our descent into Guatemala City airport; please fasten your seat belts and move your tray tables and chairs into their upright and locked positions." Anna stumbled her way, half-conscious, like an uncoordinated gorilla, to the bathroom. When she was finished, she bolted back to her seat with smiles of apology directed towards the flight attendants.

Safely in her seat, belt fastened around her waist, Anna saw lush, massive rolling mountains of green with brightly colored buildings stretching towards the clouds. Nestled in between them were roads sprayed out, resembling dried-up veins that had once brought life to abandoned cities. The runway met the descending plane like hands gently lifting to cradle them as they met the tarmac. Anna felt different somehow, as though her entire body had been holding its breath in suspension and she could exhale for the first time in a year.

She picked her way through the throngs of people moving in all directions in the small Guatemalan airport, noticing the women in traditional dresses of bright enchanting colors and feeling very much like she had stepped into another world. In her haste and anxiety, she had pre-booked a shared transfer to her hotel in the small town of Antigua but had neglected to notice the three-hour time gap. She sat in a small coffee shop where she would be meeting up with the bus group. Anna was happy to have ample time to watch Guatemalans coming and going to keep her mind occupied until then.

THE SHARED SHUTTLE van was packed with people when it arrived, but there was space in the very back corner for Anna beside a pleasant-looking man in khaki pants and a dress shirt. Anna squished her way into the van and listened to Sia on her headphones as the van careened at breakneck speed through the mountains. The jammed, heavy van made a nail-biting two-hour journey across the mountains, along narrow undulating roads to the historic city of Antigua.

Anna stepped out of the van, stumbled, and caught herself as her foot hit the cobblestone street at an odd angle. As she looked up, she saw the majestic peak of the Fuego Volcano piercing the clouds. "Wow, that's beautiful," she exclaimed, "We don't have mountains like that in Toronto!"

"Fuego Volcano is not a mountain; it burned the better part of a town a few years ago and killed over a hundred and fifty people. You can still climb up to Acatenango to see Fuego, next to it, though. People come from all over the world to see it erupting, although I'd never do it. I think they're crazy! Mas loco!" said the driver, laughing as he brought Anna's pack to her.

"Thanks for the lift and the advice; I appreciate the heads up," Anna said, laughing with him as she shoved forty Guatemalan Quetzal gently into his hand, not understanding the currency conversation or how much she'd given him.

After checking into her hotel, Anna threw her bags onto the bed and breathed a deep sigh of relief. It was as though she had been transported to another planet, and the ache of everything she had lost, although still there, had dulled even more. The further away she got on her trip, the better she felt. This was a distraction, she knew, but she needed to save herself from the depression she knew would consume her if she were to remain stagnant in Toronto. Tears, although necessary at times, could break a person if left uninterrupted for too long. Travel was a desperate lifeline for Anna. She was searching for something out in the world that she couldn't put into words, something she had lost.

Eager to explore, Anna grabbed her small daypack, left the hotel with no plan, and started walking. It was a different world indeed. Rows of beautiful, brightly colored buildings in teal, azure, crimson, and fuchsia lined the intimate streets. The bright flaxen yellow of the Santa Catalina Arch had allowed nuns to pass from the school to the convent in the seventeenth century. The arch loomed over spacious markets filled with rich tapestries that boasted the widest array of colors and patterns Anna had ever seen. She was in a world of lustrous color, and unpretentious lavishness based purely around culture and love of art.

As she continued walking, she ended up on the edge of town

and saw a pair of amiable, friendly-looking people with hiking sticks. Anna knew she looked lost and smiled at them in quiet resignation. The woman, sensing her apparent lack of direction, said warmly, "The entrance to Cerro de la Cruz is just up here, dear; it's not far. We just finished hiking up and down it ourselves." She didn't know what the woman was talking about but thanked her anyway and decided to explore in the direction they had come from. It sounded interesting to her, and they looked like locals, so she thought, 'It must be worth seeing.'

AT THE BASE of the hill, winding sets of stairs before her, Anna began to climb. After fifteen minutes of ascension, breathless, she felt immediately weightless. Where her feet made contact with the weathered ancient stone she felt only air, and it was as though the world was fading into the background entirely. She could see only the path illuminated in hazy white light from above the canopy, and the forest faded into the background. She was the only one on the path. A soft voice came to her. The voice was familiar and warm but disturbing because not only was she hearing voices but she couldn't place its origin, "Anna dear, what is it that you want? You need to follow what was lost."

"I don't want to be alone anymore," Anna screamed back into the light, balling her fists, gulping, and then forcing the air out of her lungs as she ran as fast as she could toward the apex of the path.

FAITH   FEAR   FORTUNE

The salt of sweat running into the corners of her eyes, she placed her foot on the top of the mountain, and saturated forests came into view behind the backless rows of stadium seats surrounding a simple black cross. A black and white dog lay in front of the cross, and she turned to look at Anna most peculiarly. 'What's happening to me? I'm so stressed out I'm hearing voices?" she thought fearfully.

Anna sat down in front of the cross and prayed, eyes forced shut from the sting of her sweat, tears streaming down her face. "God, can you hear me? Please show me the way. I'm frightened and I don't know where to go from here. Please help me." She could see the majestic peaks of the Acatenango and Fuego Volcanoes as the small black and white dog came to lay beside her gently. "Hey, what's your name?" Anna said out loud as she scratched the dog behind her ears. All of a sudden, Anna felt calm. She looked down at the town of Antigua, church spires rising out of clusters of single-story buildings; it was all clear and in focus.

Anna was tired and apprehensive about taking the path back down the mountain. As she began to descend, she was greeted only by friendly strangers making the trek up. "I must have been early if I didn't see anyone on the way up here. Nice to see other people here now. It's safer," she thought to herself. She turned to see the black and white dog, eyes gleaming bright, staring back at her curiously from the top of the stairs as she made her way back down the mountain.

BACK AT THE HOTEL, Anna took a quick shower and changed out of her sweaty clothing. The rest of her tour group would be meeting in a few hours for dinner, and she didn't want to turn them away by smelling like a hot mess. After a shower and a brief walkabout to explore a few more of the antiquated streets, she entered the hotel lobby and saw a group of people who looked like backpackers. They were accompanied by a vibrant man who introduced himself as Ale. Covered in intricate piercings and tattoos, he said, "Hello! Hello! Are you here for the tour?" Anna could feel the positive energy radiating from him.

"Yes, I'm Anna, nice to meet everyone. I'm from Toronto," Anna offered. A tall fellow with a black ballcap eyed Anna carefully from the back as she introduced herself to the others in the group. He was last to make introductions, "Hi, my name is Dan. I'm from Switzerland," he finally said. "You must be the pretty girl from Toronto. Are you going to be joining us for our early dinner, Mexican food?" Anna was immediately struck by his confidence, which seemed juxtaposed with his apparent youth. Dan had a dark olive complexion and a thick Spanish accent. 'I'm not a girl, kid,' Anna thought to herself.

"Yes, I'm joining you all for dinner," Anna said politely but rather put off by Dan's comments. "Great, Anna, it's so nice to meet you," piped Ale. "I wanted to also let you know that a few of the group members were asking about climbing Acatenango to get a view of the active Fuego volcano, and we're arranging a tour tomorrow,

FAITH   FEAR   FORTUNE

separately. If you want to join, it's three hundred dollars each, but you get lunch and snacks." The price seemed steep to Anna, but she remembered the stunning vista of Acatenango from Cerro de la Cruz. She decided to ignore her bus driver's trepidation and embark on this adventure that was temptingly and conveniently laid out in front of her.

9

The Temple

———

The small group of five was enjoying the warm glow of the morning sun while waiting beside the wall outside of the hotel for the tour to Acatenango Volcano to get a glimpse of the very active Fuego Volcano. The normally oppressive humidity was replaced by a mild coolness. A pickup truck arrived to take them to the base from which they would make the early morning climb. "I guess a few of us are riding in the back," Anna said nervously as she saw the cab was nearly full and climbed into the back of the pickup truck. The wind whipped the sides of their faces as Mark, Dan, and Anna sat with their backs against the cab for the winding trip.

As they approached closer to the volcano, the roads became rougher and eventually gave way to dirt pathways carved into the sides of the mountains. "Stand up and ride the mountain!" the guide shouted from the passenger seat of the cab. The three of them got up as instructed and gripped the top cage of the pickup truck, white-knuckled as they bounced over the earth up the mountain. Anna couldn't remember when she had laughed so hard as she slammed back and forth into Dan and Mark.

They stood at the base of the dormant volcano, Acatenango, which they would climb to get a perfect view of the continuously erupting active Fuego volcano. The magnitude of the tall black landmass could be seen disappearing into the clouds, and it made Anna feel small. She started having second thoughts about the monstrous climb. A week prior, she had been in the pit of depression in her condo in Toronto, staring out her window at gray skyscrapers, and now suddenly the climb of her life lay before her, taunting her shattered confidence.

Anna tensed the laces of her boots and the straps of her daypack and started walking up the mountain with the group.

HOURS PASSED as they scurried across diagonal rivers of loose black volcanic rock that snaked their way through the misty vegetation. The air grew thin, forcing the group to slow their pace. Over the crest of a hill, Anna broke from the group and ran ahead to chase a small white dog when suddenly the black silhouette of Fuego came magnificently into her view. She gasped, and the thin air came rushing into her lungs, making her dizzy.

Fuego erupted in a puff of billowing gray smoke that curled its way into the atmosphere like the hands of composers directing a symphony. Anna screamed in delight back towards the group and then started to run up the mountain away from the dense foliage to get a better view of Fuego in all her glory. As she did, everything around her faded back into the gray mist, and she saw only the black gravel of the mountain path. "What do you want, Anna? Follow what was lost," the billowing voice came again, and Anna could tell somehow that it had a female energy to it.

"I'm tired of feeling so alone! I told you this already! Follow what? I've lost everything already! Why won't you leave me alone!" she screamed in frustration, bounding up the mountain as the cold mist enveloped her body, her vision fully obscured except for the tail of the small white dog. Anna chased the dog down the path, and the mist parted to reveal Fuego with smoke now in the shape of a dog's head, fading away. Anna rubbed tears out of her eyes, and there was nothing but a giant plume of smoke now. She turned around and panicked as she realized that the group was nowhere in sight, but she wasn't alone.

The white dog, which Anna now noticed had caramel-colored floppy ears, stared curiously at her from her feet. "Hey sweet girl, there you are! Why were you running so fast? I just wanted to say hello to you." Anna sat on a grassy patch of mountain holding the small white dog in her lap, staring into the erupting volcano, feeling nothing but peace. She heard the group approaching from further down the mountain. "Anna, looks like you made a cute little friend, what's her name?" Dan said. "She's got good taste," he added with a half-smile.

FAITH   FEAR   FORTUNE

AS THE GROUP descended the mountain into the twilight, the beams of headlamps tracing stripes across the conifers along the path, Anna walked beside Dan, trading stories about their lives. Anna knew he was younger than her, yet somehow he radiated an air of wisdom far beyond his years. Dan spoke four languages and had originated from Mexico but had traveled the world. She felt an odd connection to him, and it didn't hurt that he was moderately attractive either.

"You are an amazing girl, Anna. Most women wouldn't stand up in the back of a pickup truck like that. You've got a real adventurous spirit, and I like that," Dan offered. "Thanks, Dan, over the past few months I've started to learn that life is too short and you need to do things that scare you because you might not get the chance again. Sometimes you just need to breathe, hold on tight, trust, and see what happens, I guess," Anna replied as they rode back to the hotel.

In the back of her mind, all she could think about was that strange female voice she heard at the peak of Acatenango, and the top of Cerro de la Cruz. 'What is happening to me? Was that just the lack of oxygen, or the altitude? Hearing voices is one of the more cliché indicators that something isn't quite right,' Anna thought to herself. She didn't like the label 'crazy' because it had been used for decades to abuse people who had a mental health condition, but she did wonder about her sanity.

By the time they all got back to the hotel, it was well past ten and they were all exhausted. Restaurants were shuttered for the night, so they passed around fruit and some granola bars. This was quite alright as far as Anna was concerned. All she could manage was a quick shower before rolling herself into the soft, cozy bed. She was asleep within five minutes, dreaming of mountains and volcanoes covered in frolicking wild dogs.

THE NEXT MORNING, her alarm blared, and in a daze, still exhausted, Anna grabbed her backpack and was out the door to catch the bus at a quarter past six. As she lay back in the tour bus, she witnessed life's theater like the frames of a blurred film. Tanned, slim villagers on dilapidated bicycles moved past every different kind of makeshift establishment. People were selling fruits, beer, groceries, clothing, and housewares from places with hand-painted signage in bright but peeling block Spanish letters. Children and dogs rolled through the streets like tumbleweeds, unaccompanied and liberated from the watchful eyes of their families. Guatemala was free, lively, and culturally rich, and the tour group was on its way to its heart, the ancient ruins of Tikal.

Anna loved traveling with a group but had only done it a few times when she was younger. It was always a great way to meet people from all around the world, from different cultures, backgrounds, and walks of life. This group was no exception, boasting explorers from Australia, the UK, Spain, Portugal, Italy, Switzerland, and of course

Canada. The smaller group of volcano trekkers had joined up with the rest of the group on the bus and were now a healthy group of twelve, including Ale, their local guide. "It's going to be hot out there today, guys, so make sure you've got your water, your hat, and sunblock," Ale cautioned.

WHEN THEY ARRIVED at Tikal, the group piled out of the bus and proceeded to walk to the temples. They were greeted by a raccoon-like creature called a Coatis, who didn't seem apprehensive about getting right up next to them as he scavenged for food. Anna had never seen anything like the striped-tailed mischievous-looking critter. Weaving through the ancient buildings covered in hundred-year-old moss, the group of twelve arrived at the temple of the two-headed serpent. Anna could feel the spiritual weight and the energy of generations who had stood in the shadows of the temple, gazing up in awe at its splendor.

Standing right next to the two-hundred-and-thirty-foot stepped structure, as forty-five-degree heat beat down on her head, Anna placed her hands on the temple steps just above her head and pulled herself up onto the first huge block stone step. "Is anyone else coming with me?" she called back to the group as she saw them all walking away.

"No, you can go. It's too hot, Anna. Maybe we will try later in the afternoon when the sun is a bit lower in the sky. Meet us at the picnic tables for lunch in thirty minutes," Ale called back. Anna kept climbing, and as she climbed, she noticed that the limestone was surprisingly cool to the touch. The ancient stairs were starting to crumble away, and she was sure to carefully place her feet on each step, slowly and steadily moving up the temple.

The sun was bearing down, and she was sweating profusely. As she got to the third section of the temple, she looked up and realized there was no one else up there except for something small and brown that seemed to be curled up in the doorway at the top. Determined to reach the top and see if the creature in the doorway was alright, Anna started climbing steadily again. A monarch butterfly glided elegantly from the trees just above her head, the striking orange and black wings contrasting against the white limestone.

She kept climbing, and the steep incline of the steps and the height started to make her dizzy, no shade, just the incessant sun. Everything around the temple was suddenly undulating like a mirage in the heat, and through the salty sweat that dripped into her eyes, all she could see was the final section of the temple crisply in focus. Where she thought she had seen an old wooden door from the bottom of the temple, was now a gilded shining one, and that voice again, "Anna what are you looking for? Follow what was lost." As she neared the top, she realized that the brown animal was a little dog that looked like a terrier.

"Voice! Who are you, and why do you keep asking me that? What are you talking about?" Anna cried, "What do you want? What's happening to me? I'm losing my mind, aren't I? That must be it!" Anna continued in desperate pleading.

The voice came back clearer, "You're not losing your mind; it's clearer than it's ever been. I think you know who this is, Anna. I think you've known since I started talking to you on Cerro de la Cruz. You haven't lost everything; you need to follow what was lost, just trust and let go."

"I don't know who this is! What do I need to trust? Everything is gone; I've lost my partner, my memory, my ability to work. I'm scared I'm losing my mind and worst of all I lost my best friend in the entire world, the one who was keeping me going and getting me through the days. I don't know how to do this without my dear Ella. I want to die," Anna wailed.

"Bingo," the voice said, "You haven't lost me at all. Anna, I'm nearer to you than I ever was before and I'll remain here with you as long as you keep seeking what was lost. I can show you the way now because I'm free," the voice continued, "I can tell you need a little help getting where you need to go right now, but I can assure you Anna, that death isn't what you think it is and you need to follow what was lost to find out."

ANNA FELL to the ground in front of the temple of the two-headed snake and let the tears of elation soak her cheeks, "Is it true?

Can it really be possible that Ella hasn't left me? Ella is that you?" she cried, but the voice had vanished. As the tall trees surrounding the temple came back into focus and grew lush and viscerally green, the small brown terrier came to lick the stream of tears that had fallen onto her legs and stared knowingly into her eyes.

10

The Boy

When Anna woke up the next morning, she lay in bed, contemplating what had happened at the temple of the two-headed serpent. She seriously considered whether her experiences had been part of some sort of a dream. Yet, she hadn't passed out, and they felt so real to her, so palpable. Although she hadn't been to church regularly in recent years, Anna did believe in God and considered herself a spiritual person who had faith in things that couldn't always be felt or touched in a tangible sense. She didn't fully understand what God was, but she felt deep in her that all of life was not by chance.

Anna always had a tough time picking one religion over the other because they just seemed like interpretations of the same magnificent force that bound us all together in spirit. She was baptized Anglican as a child and attended Sunday school until she was twelve. Afterward, she studied and later identified with, Buddhism as a teenager. Anna had gone to a Catholic Church for a few years with a boyfriend, and most recently had been a member of a Missionary Christian Group. The truth was that she loved all religions because, at their core, they were about the hope we all have of a continuation of the spirit after our physical bodies give up.

Anna felt with every cell of her body, and the spaces in between, that the soul cannot be destroyed even in death. It was true that the fuzzy feeling of warmth that she had felt constantly at her side that brought her moments of pure presence and pause amid anguish, was the spirit of her dear Ella. There was no doubt in Anna's mind that it had to be true now, and Ella was showing her the way and guiding her even after she had left her physical body.

Encouraged by the deep certainty that she was not making her journey alone, Anna rose and got ready to face the day. The tour group was going to be making their way to Caye Caulker in Belize, and it was promised by Ale to be a long journey involving buses, ferries, and taxis.

SEATED COMFORTABLY in the first row, Anna watched as the tour group members funneled into the bus. "Can I sit here Anna?" Dan asked as he sat down beside her. "Absolutely not," she replied, joking as she laughed at his question which was more like a statement.

"So, a few of us were interested in checking out the local scene in town and Ale is going to take a small group of us out to some local bars later on, after the group dinner, are you up for joining us? I would like it if you came with us," Dan stared at Anna as he brushed her knee gently with his hand, his large eyes doe-like, his accent thick and low. 'How old is he really?' Anna thought to herself.

"Yeah, I'm totally up for checking out some local bars, and I'm sure Ale knows all the best ones. Maybe we'll get some dancing in," Anna agreed. She was really looking forward to the evening now, and as they talked about their families and lives back home, she was starting to develop what she thought might be feelings for Dan. This utterly terrified her. The last time she developed feelings, it didn't end well for anyone, but she tried to focus on the conversation and push those fearful thoughts out of her mind.

FIVE HOURS and four different types of transport later the group had at long last arrived in Caye Caulker. They finally finished dinner at a seafood restaurant at nine in the evening and were splitting off to go to different spots in the city. A few were going for gelato, another cohort decided to go back and get a good night's rest, and five of them, Anna, Dan, Ale, Mikaela, and Robyn, were going to check out some local salsa bars. "Ok, let's do this gang," Ale whooped, rousing them as they headed off in the direction of the bars.

Alive with sound and color, laughter and movement, the salsa clubs were a smorgasbord of revelers. As Anna danced and spun, a flamboyant man in a shiny white suit and a neon pink dress shirt took off his star-shaped sunglasses and placed them on Anna, smiling while dancing around her and wildly waving his arms. She couldn't help laughing and caught sight of Dan on the edge of the room who was standing with Ale while eyeing her intently. Gesturing Dan over to join, Anna grabbed his hand and pulled him into the chaos of the packed dance floor. When the crowd finally started to thin out hours later, Anna realized the rest of the group had left.

AS ANNA and Dan walked back to the hotel, tipsy from dancing, and intoxicated from the loud music and cocktails, they were alone on the sandy white beach streets. Anna clumsily stumbled over a gaping hole in their path, grabbed Dan's hand out of instinct, and gingerly jumped up on the curb laughing. "Can I kiss you, Anna?" Dan asked her matter-of-factly as he leaned in and kissed her. Anna briefly responded but caught herself and paused, "Do you know how old I am? How old are you anyway?"

"Age doesn't matter to me, Anna. I know how old you are because I asked Ale already. I like you, Anna," Dan reassured her. Anna was hesitant, "I don't want to get hurt; I gave you the rundown on what happened to me last year, right? I don't have the emotional capacity to get my heart ripped out again, Dan, I just don't. This isn't a good idea."

"It's okay, Anna, I won't hurt you. I like you! You're the nicest, warmest person I've met, and to top it off, you're beautiful. You're perfect, Anna," Dan reassured her again as he kissed her and embraced her strongly, pulling her into his arms. Anna felt safe and content with Dan, but a nagging fear was crouched in the back of her mind. She tried to push the fear down deeper, into the recesses and silence it. She wanted to feel loved and to love back, and she didn't want to feel alone anymore.

Anna embraced Dan and kissed him, replying, "I think you're wonderful, Dan. I'm so glad that we met. I'm grateful for the nice things that you said to me, but please don't call me perfect because I'm not anywhere near perfect." For that moment, she thought better of her fears and recalled one of her favorite quotes by Henry David Thoreau, who said, 'There is no remedy to love but to love more,' and this gave her comfort.

During the rest of the walk back to the hotel, Anna felt weightless and like a child again. Floating through the sandy streets, everything seemed to take on a bright beauty and charm. Was she able to fall for someone again, and could she actually let herself feel this way? The trepidation was giving way to hope that perhaps her love life wasn't over at all and that something wonderful might be on the horizon.

THE NEXT MORNING, Anna woke up on cloud nine. 'Ella is it true, did you lead me to this place to find the love I've been wanting for my entire life? Is this what you've been trying to tell me?' Anna thought to herself that maybe everything was working out the way that it was destined to and that Ella was guiding her. 'Follow what was lost right? Well, I lost love, so maybe this is what she meant," she thought, 'Perhaps all of the terrible things that had happened in my life were leading me here to this very place to meet Dan.'

Anna sprung from the bedsheets like a flower freeing itself from the bud. She felt alive and full of hope for the future as the early morning sun shone fiercely through the hotel window. Throwing on a pair of shorts and flip-flops, she walked in the direction of the ocean, two blocks from the hotel. She could hear the ferocious crashing of the waves growing louder. The rest of the group was still asleep, 'It's a shame they are missing out on the beauty of the sunrise,' she thought.

Sitting on the edge of the ocean, it was as though she was on the precipice of her limitless future, feet dangling with toes gracing the vast endless salty waters. Like her future, the great depths were mysterious and held the wonders of her imagination. It felt in that moment that she was surrounded by an intense beauty and nature's one true connected spirit. It enveloped and overwhelmed her, bringing tears to her eyes. 'Everything painful is behind me, and Ella is with me guiding me still. There is so much hope in this place,' she thought quietly.

THE GROUP JOINED together for breakfast back at the hotel an hour later. Anna noticed that as she smiled at Dan from across the table, he was shifty and wouldn't make eye contact with her. 'Maybe I'm just reading too much into things,' she thought reassuringly to herself, 'I'm still a bit sensitive from everything that's happened. Ella told me to trust and let go and to seek what was lost. I lost love and I need to trust the universe to find it again.'

FAITH   FEAR   FORTUNE

The itinerary called for a free day to explore Caye Caulker and a few of the group members were going to make their way down to the end of the cay to lounge at one of the famous sunken in-ocean watering holes for a few drinks in the sunshine. As they walked, Anna was hanging near the back of the group and went to grab Dan's hand, smiling at him. He pulled it away abruptly. "Can we please not do that in front of the group, Anna," he whispered to her, "I don't want to make anyone uncomfortable."

Assuming that he was just shy with public displays of affection, she didn't think much of it, nodded back, and continued to walk alongside him toward the gap. They spent a beautiful, relaxed day submerged in the salty turquoise waters of the gap, sipping cold drinks in the hot sunshine around circular coconut tree tables. On more than one occasion, Dan had run his leg against hers playfully in the water, stealing glances and smiling.

AT THE END of the day, feeling totally happy, Anna walked back with the group, drenched in salty water from both her sweat and the ocean. Dan seemed distant to her as they walked, and she started to feel that his energy had shifted again. As they neared the hotel Dan pulled her aside, "Listen, Anna, I like you, and I was telling my best friend about you last night and he was so happy for me. However, I talked to my mom this morning and told her about you, and how old you are, and she was really upset. My parents want me to have a traditional family. I don't think I can do this with you, Anna, because of your age. I hope you understand."

Anna couldn't speak, and it felt as if the air, blood, and life had been knocked out of her body. The weight of his words came crashing down on her head like an anvil, and she could only stare in suspended horror. The best she could manage was, "Okay. That's no problem. I'm going to go for a walk now before dinner, see you later." She walked quickly around the corner and out of sight and then started to bolt toward the ocean. She passed the familiar spot where she had witnessed the sunrise in the morning, tears streaming down her face.

BY THE TIME she had reached the break wall, the tears were gushing like a faucet, her stomach was in knots and all she could see was the ocean ahead of her. Anna's salty tears mixed with the salty seawater as she hung her head in the dusk of the evening and cried, feeling utterly defeated. Anna felt stupid for trusting again so easily, and being so open with her heart. Ella's voice came to her, "Anna, that's not where you're meant to go. I'm here to help you get where you need to go. Follow what was lost."

"Everything I do goes wrong, El. You were the only good thing in my life and I don't know why I even try anymore. What's the point in anything without you here, El? I miss you so much. Why am I even alive? It hurts so much I can't breathe. No one else loved me like you, and I don't think anyone ever will," Anna pleaded into the darkening abyss of the ocean, towards the soft familiar voice.

Ella's voice came back again, not human but more like a voice all around Anna that she could not control, "Yes, Anna, I love you so much still and I never left you. I could never leave you because our love was pure and good, full of spirit. Like I said before, I will make sure you get where you need to go on this journey, keep remembering to follow what was—"

"—Lost. I know, El. Please tell me what this means. I don't understand. I thought I understood, but I am really confused and lost. I'm trying to understand what you want me to do and where you want me to go," Anna said, exhausted through tears as the waves crashed at her feet, soaking the ends of her shorts.

"Do you remember when we would go for walks in the park and then you would throw my ball for me as far as you could across the dewy morning grass? I would leap joyfully towards the ball not knowing where I would end up, these moments of abandon were truly the happiest moments in my life with you, Anna. I enjoyed being a dog because of you," Ella came through softly.

"What are you now, Ella?" Anna started to settle down and grew more curious than anything else.

"That's not the point, Anna. The point is that you need to trust and let go. Stop holding on so tight to things that don't belong to you. If you follow what was lost, what is meant for you will find its way to you. I'll make sure you get there, for now just keep moving forward," Ella said with absolute knowing.

The lights of the bars and restaurants along the coast grew brighter and lit up the sides of the harbor again. Anna had almost forgotten about Dan, but the pang of hurt struck her insides. 'I have to keep moving forward,' she thought to herself, 'I don't know how I'm going to get through the rest of the tour, facing him, but I need to keep going for myself and Ella."

The next morning the group was bound for Playa Del Carmen, Mexico. Every time Anna looked over at Dan, laughing like nothing had happened, she felt betrayed. Throughout the last leg of the journey, after they boarded the bus from the Belize-Mexico border, Anna had been forced to sit behind Dan. Listening to him joke and laugh while she was in tears was too much for Anna, and she decided she would strike out on her own following the next stop, leaving the tour. 'Keep going, Anna. Follow what was lost,' Ella's words etched in her mind.

Thanking Ale and explaining the situation, Anna booked a flight to Mexico City from Cozumel. She decided to plan where to go from there based on the available tours on sale and just trust and let go as Ella had said. The first tour that came up, leaving on the right dates, was Costa Rica. 'Okay, Ella, we're booking a flight to San Jose, and we're going to see some sloths. I hope this is the right move,' Anna thought as she pulled up an app on her phone and booked her flights.

FAITH   FEAR   FORTUNE

Manuel Antonio National Park, Costa Rica

11
The Leap

Anna arrived in San Jose late at night. It had taken her well over twenty-four hours to reach Costa Rica from Cozumel, and this included layovers and an overnight flight from Mexico City. The city lights twinkled as she sat in the back of the taxi bound for the city center, and she couldn't wait for the warm softness of a bed to quell her exhaustion. Traveling and being in an airport felt like a form of meditation to Anna—consumed with noise and activity, there was little time to think about anything other than getting where she needed to be. Now, however, in this quiet space, her thoughts were taking over.

'It was always the idle moments that brought a man or woman to ruin,' she thought she had heard that somewhere before but was too tired to remember where. 'Here I am drifting aimlessly, alone in Central America, having done zero research on any of the places I'm going. I have no idea whether this country is even safe.' Anna's stream of sleep-deprived negative thoughts were punctuated by red streetlights as they neared downtown and the hotel she had booked last-minute. By midnight, Anna had arrived and fell into bed, still clothed, and passed out immediately.

ANNA WAS EXCITED to meet up with the new tour group in Costa Rica the next evening. There was a father and son from Vancouver, an American, several Germans, and British and Irish nationals eager for the pursuit of adventure within Costa Rica's rich biodiverse beauty. In trusting and letting go as Ella had said, Anna had arrived in one of the most beautiful places in the natural world. Perhaps this was where she needed to be.

The plethora of activities available to Anna was sure to challenge her fears and provide the much-needed distraction she was looking for to get her mind off Dan's betrayal. She was now, after all, literally miles away from him and that terrible experience and ready to tackle new adventures. 'Follow what was lost Anna,' she thought to herself. What did that even mean? Her confidence was certainly shattered to pieces. Perhaps doing things that terrified her would bring her to where she needed to be.

Alex, their guide, said that the group would have the opportunity to participate in riding the longest zip line in the world, followed by an optional bungee jump at the end. Having had a mortal fear of heights even before she had seen Calvin plummet to his death from her balcony, she decided that conquering this fear was what Ella must have been talking about. It was time to face her fears.

STRAPPED INTO the harness in what was called their 'Superwoman' pose, face down towards the lush emerald jungles of Costa Rica at the cruising altitude of a small aircraft, Anna would be zipping across at fifty kilometers an hour with only a thin wire separating her from life and death. If this wasn't facing her fears and chasing her lost shattered confidence, she didn't know what was.

Taking shaky steps up the perforated metal stairs to the tall platform where she would be strapped in, Anna's mind was blank. By the time she had reached the top where two strangers were waiting to connect her flimsy harness to the wire, her brain had somehow switched on, and she was in full panic mode, palms sweating and heart racing. 'What the heck am I doing here? I'm going to die in this place, I must be out of my mind. Oh, El, please tell me that I'm doing the right thing.'

The strap tethering Anna to the platform was released, and she started to plummet toward the sprawling canopy of dense, lush rainforest that was interspersed with houses below. The roofs of homes were like specks dotted throughout the greenery. "I'm flying! Ella, I'm a bird," she screamed into the tall forests as the wind whipped her face. All Anna could see was the zip line cutting a path through the sky above her as she soared like an eagle.

"I'm getting you where you need to go, my Anna. Enjoy the journey, trust, and let go. Keep following what was lost. You're well on your way," Ella's sweet voice rang in her ear in a harmony that's beauty matched the flight of abandon she was taking, "Keep going, Anna."

AFTER THE FLIGHT through the canopy, Anna was allowed to jump from a twenty-story platform that was connected to a flexible cord dangling casually over the unforgiving ground. Never having done anything that challenged her fears so directly before, Anna watched as the muscled man ahead of her walked up to the edge of the platform and stood motionless for ten minutes. The longer the man stood, the larger and firmer the knot in her belly became.

Anna felt like she might be sick as she saw the man walking back towards her, exclaiming, "No way am I doing that shit, man," in resignation. Could she do it after watching someone so close to her fall from her balcony onto the concrete, to their death, if that muscled man couldn't even do it? she thought nervously, 'I must be twice his age! How can I do this?' She started having flashbacks of Calvin's dead body on the concrete as she edged along the narrow walkway that shook back and forth every time she took a step. She had to steel and focus her nerves every time she caught a glimpse of the ground below through the wide-gapped metal grating.

Edging further to the end of the platform, with no more places to put her feet, her legs began to wobble like jelly, and it seemed like they might give out before she could take another step. Gazing down at the group, the minuscule tops of people's heads surrounding the giant target that had been painted on the ground finally made her topple to her knees, "I can't do this, I can't! This is just too much, I'm trying here but this is just too high," Anna started to exclaim as her entire body recoiled in instinctual fear.

"You can do it, and you will do it. The more you think about it, the more you cheat yourself out of the experience of doing it. Trust me, I know," the attendant said from beside her. Anna looked up towards the sun. "El! Ok, I'm doing it El, for you!" she screamed as she stared up at the bright blue cloudless sky and thrust her entire body into it all at once, trusting the universe, trusting it all. As she plummeted to the earth, the fear fell out of her body and her mind like a rush of blood escaping the crown of her head and setting her spirit free.

Down on the ground when she had been removed from the harness, she broke down in tears in front of the onlooking strangers and her tour group, who were watching the jumpers in awe, "I did it!" She screamed, "I'm not afraid anymore. I did what I didn't think I could ever do, in a million years. That was epic!"

FOR THE REST of the time that she was in Costa Rica, Anna said yes to everything that was put in front of her, she started saying yes to life again. Whitewater rafting, waterfall rappelling, sleeping in tents in the jungle, and hunting for orange-kneed tarantulas in the black damp midnight rainforests, she said yes to it all. A crippling fear had been hanging like a haze of confusion, causing inaction, but it was fading away. In fear's place was something Anna thought she had long lost, her confidence in herself.

WHEN THE TWO-WEEK journey through the small Central American country was over, the fear started taking hold of Anna again. The thought of going back to the empty condo where she had endured so much combined with her newfound confidence, emboldened her to continue on her travels. Ella had, after all, kept telling her to keep going.

Anna had always considered herself to be a relatively sporty woman, and she had kept herself in decent shape, so she decided she would find an active vacation as a continuation of all of the rich activities she had been introduced to in Costa Rica. Trusting, and letting go, she prayed, 'Show me the way El, and tell me where I need to go.'

Anna set her intentions to honor Ella's wishes as she pulled up a list of tour packages on sale. The first one that came up was an adventure tour to 'Columbia and The Lost City'. "Holy shit," Anna said out loud, "OK, I'll follow what was lost, this hike shouldn't be too hard," she continued taking out her credit card and booking the tour, which was leaving in seven days from Santa Marta.

12

The Elder

Piecing together her way from San Jose to Colombia was relatively easy for Anna. She quickly found a low-cost flight with a fast connection through Bogotá that was leaving the very next day. Anna wouldn't be particularly upset leaving San Jose. San Jose, unlike the rest of Costa Rica, was loud, dirty, and dangerous. When she had arrived in the city two weeks prior, it had seemed purely menacing at night. She had even noticed a man following her on one occasion and had sought safety in a small, well-lit corner store.

Garbage lined the streets of San Jose, which were, for the most part, devoid of tourists or inhabitants. Having worked in big cities all over the United States in her twenties as a consultant journalist, Anna knew that the surest sign that a city was unsafe was when it was deserted after dusk. A lack of watchful eyes left criminals to go on about their sordid activities, unchecked through the twilight hours. On the cab ride to the hotel, Anna had seen someone smash glass bottles into a window and two men fighting in the middle of the street, obviously intoxicated, with no police in sight.

FAITH   FEAR   FORTUNE

Anna was glad to have seen what she did early on in her stay as she was entering the city because it encouraged her to take extra precautions with her cash and passport while walking around during the day and especially at night. When she had just arrived and had been walking through the San Jose Craft Market at midday, she had felt a tug at her backpack and spun around to see a man walking briskly away, disappearing behind one of the stalls. She was certain she had gotten lucky and averted a near theft.

Anna wasn't afraid of the San Jose thieves after that. She was smarter and carried her cash close to her front in a waist pack, walking with her head held high and a projected confidence that seemed to keep the unsavory characters at bay.

AFTER A DINNER of fish tacos, Anna strolled back to the hotel and settled in for her last night in San Jose, excited for the ensuing journey to Colombia and the adventures that awaited her. Faith had replaced hopelessness and depression, confidence had replaced fear and inaction, and who knows what fortunes would smile down on her tomorrow. She rolled over gently and went to sleep.

The flight was leaving early in the morning, and Anna had packed and gotten ready the night before so she could maximize her hours of sleep. She sprang up at four am, showered, threw on a set of clean clothing, and was out the door in fifteen minutes. As she boarded the flight to Bogotá, she felt as free as the brightly colored parrots she had seen gliding across the rooftops of Guatemala. Anna lightly placed her carry-on in the overhead compartment with ease, gently sinking back into a shallow but restful sleep.

CONNECTING IN BOGOTÁ was a puzzle to be solved. Making her way out of the international terminal over to the domestic one was more of a maze than other airports for Anna, with her limited command of Spanish and lack of cell service, but she did it, if only through trial and error. From Bogotá, it was a small hop back up to Cartagena where she would meet the tour group. They would be traveling together to Santa Marta to begin The Lost City trek, which was world-famous for being the widely touted 'Best hike in South America'.

AS HER TAXI wove its way through the old walled city of Cartagena, Anna entered a traditional festival of music and dance and had to disembark two blocks from her hotel to make the rest of the trip on foot. With perfect ebony skin and soft complexions, women in voluptuous Spanish dresses twirled apart and together in a choreographed dance. Men with colorful embroidered sashes around their waists and large-brimmed straw hats smiled widely as they gleefully stomped their feet to the beat of rhythmic drums. Anna felt alive and consumed by their beauty, and their joy was infectious.

By the time she had reached her hotel, Anna felt energized, and she was ready to explore the beautiful city. Her base was a quaint, historic hotel in the heart of the walled city, with rustic acacia wood features everywhere. A quaint pool in the center was adorned by lush vines and flowering climbing greenery. Anna was given a room overlooking the pool with old-style coffee plantation shutters and large potted palms framing them. Throwing her bag down, she felt an urgency to make her way back outside and join the festivities. Life was beckoning her.

Pulling the metal latch across the front gate that separated the hotel from the outside carnival, Anna burst into the street and joined the parade of revelers, her money belt squarely across her front, of course. Stomping her sandals to the beat on the pavement and looking very much like a misplaced tourist gringo, Anna didn't care. She threw her hands wildly into the air with the reckless abandon she had seen in the clubs in Caye Caulker and danced her heart out.

Anna followed the parade through the streets of the old walled city of Cartagena and considered it to be possibly the best free walking tour of the city that could have been orchestrated. Ancient Spanish architectural influence was evident everywhere she looked. Cartagena felt ancient not simply for tourist purposes, but genuinely deeply, and authentically. Women with fruit baskets balanced delicately on their heads, horse-drawn carriages, boys selling mango slices, and churches filled with people from all walks of life. It had an innocence to it that seemed to be insulated from the modern world.

This, Anna decided to herself, would be a great base from which to start her Colombian adventures. She parted ways with the carnival at the Raphael Del Castillo building of bright yellow and intricate ornamental facade, and meandered down a side street, under a canopy of bushy fuchsia flowers. She stopped at a bright pink bicycle against a robin's egg blue store with lemon-yellow trimmed windows, feeling like she was inside a vivid painting. Truly Cartagena was a magical place.

On the street was a small shop with a wooden hanging sign that read 'Gelato' in peeling, fading white letters. Anna entered the store and eyed the luscious concoctions, each one ornamented with decorations of its flavor. 'Maracuya', Anna didn't know what that word meant, but there was a large, deliciously ripe passion fruit halved on top, so she decided on it. Enjoying her gelato in Plaza de la Paz, she felt content to sit and watch the theater of tourists and buskers and couldn't wait to meet her hiking group in the evening.

SIX ROLLED AROUND quickly, and Anna, sitting in the lobby of the hotel early as usual, saw a couple of people who looked very much like hikers, with their sturdy boots and slim physiques. "Hi, are you guys here for the tour?" she reached out.

"Yes, I'm Yves, from Montreal, Canada," the first guy answered. "I'm Priya from Quebec, also in Canada!" the girl answered enthusiastically, "Are you doing The Lost City trek too? I'm so nervous about walking sixty-five kilometers, but it will be a good challenge!"

'The website said it was forty-two kilometers.' Anna thought nervously and wondered if her face betrayed her. "I'm Anna, from Toronto! Nice to see that there are a lot of Canadians here. Yes, I'm doing The Lost City portion of the trek too. I'm excited about it but also nervous like you, Priya," Anna replied, "Do you know how many people are in our group? Have you met anyone else yet?"

"Yes, actually, I met our guide already when I came in early yesterday, and he said that we're a group of nine people. We're all meeting upstairs on the seventh floor for orientation in half an hour to go over the route and what to expect, so we'll meet everyone there," Yves replied.

AT SEVEN, Anna met the rest of the group. In addition to the Canadians, Yves, and Priya, there were six others as well as the guide, Hugo. Polly and Carol were friends in their early twenties, from Germany. Mel, Joe, and Peter were all solo travelers in their late twenties and early thirties, from the United States. The last member of the group was Nelly from London who had recently graduated college and was backpacking through South America before she would begin her career in medicine.

'Everyone here looks quite young. I'm the oldest one by a long shot, and I'm having my birthday on the trip so I'm going to be even older by the time it's through.' Anna thought quietly as she introduced herself to the hiking circle, "Hi everyone. I'm Anna from Toronto, Canada, and I'm excited to be joining you all! My fun fact is that my ancestors were related to Mary Shelley, so I guess I'm related to her too. I do warn you all, I'll probably be at the back of the pack but I don't expect anyone to wait for me."

"That's fine Anna!" Hugo piped up, "It's not about how fast you're going, everyone should feel comfortable hiking at their own pace. It's not a race and it's going to be very challenging at times, so I'd prefer you take your time. It will be much more challenging than the hikes you're used to, I'm sure! I've been leading groups for ten years and I've heard over and over from our seasoned hikers that it's one of their toughest and best hikes. I can promise you all that you will make memories that will last you a lifetime!"

'The hikes I'm used to? I think the last time I hiked was ten years ago on a gravel path through the park,' Anna thought. She had picked up a set of walking sticks and a sturdy pair of boots, which she had read in online forums were essential items to take for the hike to The Lost City. Anna had never actually used a set of walking sticks or hiking poles, as seasoned hikers refer to them, but all of the blogs listed them as a must-have, and with her limited hiking experience, she wanted to be as well-equipped as possible.

"Alright crew, we're going to go with the team name suggested in our WhatsApp Group, 'Pumas'! Team Pumas! Tonight we're going to walk over to Getsemani to have some dinner at a traditional restaurant that serves authentic Colombian dishes. It's one of my favorite places. Getsemani is a very artsy place and there are a lot of different options in the restaurant for all dietary restrictions. Bring your appetite because the food is really good, and so is the coconut lemonade," Hugo finished, "Let's go!"

THE GROUP WALKED through La Paz square, across the park, and wove its way through Getsemani, which was adorned with beautiful and varied styles of street art. 'This isn't graffiti; it's a free gallery of masterpieces,' Anna thought as she walked along. There were Colombian women with every interpretation of the Afro imaginable. Colorful and striking airbrushed female faces stared confidently and challengingly from the flat parts of the cracking walls of bistros in narrow alleyways, in strong defiance of passers-by.

ANNA ENJOYED A DINNER of cheese and seafood arepas, and two large glasses of coconut lemonade that were deliciously sweet and thirst-quenching. The next morning the group would be traveling to visit an authentic Arhuaco Native community to learn more about their ancient traditions and culture. Anna was excited that they would get a chance to speak to one of their elders.

She had a lot of questions for him and was eager to hear his wisdom. She fell asleep dreaming of the Colombian jungles full of exotic and beautiful people.

IN THE EARLY morning hours, Anna rose and got ready quickly. At five-thirty, the bus was ready to go in front of the hotel and the group would be checking out and making the long, four-hour coastal drive from Cartagena to Santa Marta. Anna fell asleep again in the bus, listening to soft music, and woke up hours later when the bus arrived at the Arhuaco Native community mid-way through their journey.

A pair of guides met their group at the overgrown trailhead, one woman with prominent Arhuaco features was wearing modern sportswear and a pair of crisp new Nikes, while the other one was dressed head to toe in a traditional stark white tunic and was weaving a intricate mochila bag. "Make sure you're wearing an adequate amount of mosquito repellant because the bugs are feisty here," Hugo cautioned the group before they set out down the narrow winding dirt trail bordered by thick encroaching vegetation.

The Arhuaco tribe, being one of the few untouched tribes left in Colombia, was eager and proud to showcase their traditions, inviting the group into the village. As Anna and the group hiked over the last hill they saw the peaks of the circular adobe homes, capped in straw, sprawling out across the village. Descending in between the houses,

Anna was struck by a small building that seemed different and had no windows. It was made of huge stones, gray and ominous looking. "What is that building? Is it a house?" she asked. Hugo translated Anna's question for the villager woman, whose name was Bolsilla.

"That's a prison, and she told me that there has been a man in there for twenty-two years. He has never left the building except maybe a few times, for a few minutes each time. She said the only person he is ever allowed to talk to is the village elder, for advice on his path to rehabilitation. He is there because he murdered one of his fellow tribesmen," Hugo translated swiftly and dolefully.

Anna couldn't believe that someone could be locked away in a small ten-foot square building in the forty-degree heat with no windows and survive that long. No matter the crime, it still seemed to be a very extreme form of isolation, torture even. Anna wondered if the man could hear their foreign voices from inside the clay-walled prison and hoped that if he could, it at least provided him with some comfort or distraction.

AFTER THE VILLAGE tour, the group arrived at the home of one of the elders on the edge of the village. Chickens, dogs, and young toddlers ran around the exterior of the large circular home in every different direction. A puppy came up to lick Anna's leg, and she lifted him gently, kissed him on the forehead, and put him on her lap where he went to sleep. The group sat in a semi-circle around an intricately carved wooden bench with a handmade hammock in front of it. In the hammock sat a man who appeared to be in his late seventies, his face worn from the sun. The leather-faced man, who turned out to be the village elder named Izquierda, was wearing a white cap that contrasted greatly against his dark coffee-colored skin.

"This is Izquierda, the village elder. People from all over the village come to get advice from him, and he maintains the rule of law in the village. He also decides what the appropriate punishments will be for any crimes committed in the village. You're welcome to ask him any questions you like," Hugo translated. 'I guess he's the one who locked the man up for twenty-two years; I better not upset him by asking stupid questions,' Anna thought to herself.

THE GROUP INQUIRED about his culture, and the history of the Arhuaco people, and he answered gladly yet solemnly as he held his Poporo from which he chewed a strange white powder and spit it out beside him. Izquierda explained that the device functions much like a set of worry beads, and he writes down all of his troubles into it. They are encoded there forever, leaving him free of distress. 'I wish I had one of those. That's like their version of a diary,' Anna thought to herself.

"I have a question for the village elder," Anna posed her question, "I'd like to know what his advice would be for someone who is completely disillusioned by the modern world and has experienced only pain through it." The question was translated to the elder who thought hard for a long while and nodded in Anna's direction pensively. He gave her the first knowing smile she had seen since the group sat down. He spoke for what seemed like five minutes to Hugo, who was translating for the group, taking long pauses in between to carefully consider what he would say next.

Hugo attempted to translate the answer as best he could, "His message to anyone feeling this way, Anna, is that they need to find a way to better connect with nature and the natural spiritual world. We are a part of nature, and not separate from it. Modern life separates us from our one true nature, and the pain felt by that separation can be difficult to reconcile and manage. Fixing this dilemma is about getting back to the spirit essence. You are the natural world, and it is you. He said that everything is one. He also wanted me to tell you that you have an animal spirit guide close to you, Anna. Her divine female energy is a part of you, and you should listen to it carefully."

"Please tell him how grateful I am for that wisdom, and for offering his time to us. I will do my best to put what he is saying into practice in my life. Tell him I feel the presence of my spirit guide and I am grateful to him for confirming what I feel in my heart," Anna said in gratitude, smiling at the elder and nodding in deep respect and thanks. She was also so grateful for the acknowledgment that Ella was indeed with her, and she was not imagining things. It was all real.

WALKING AWAY from the village and making the hike back to the bus, Anna couldn't stop thinking about what the elder Izquierda had said to her. It was true, her life had a certain momentum in it up until Calvin's suicide had brought it to a screeching halt. She never really had a chance to sit back and think about the choices she had made in life, and as a result, Anna had found herself as far away from the natural world as possible. She was locked into a life of concrete, in the city in the quest for monetary fortune, and had largely been on autopilot for the past twenty years. She didn't even know what was important to her anymore.

Everything that had happened to Anna, however negative it had been, was making space in her life so she could figure out what she needed to be happy. The elder Izquierda's advice resonated loudly within her on the deepest level. Anna needed to get back to nature and the natural world and was glad to be venturing out on The Lost City trek. The hike would be five days immersed in the deepest jungles, amongst outlying tribes in the remotest possible location. "Follow what was lost, Anna," Ella's voice came through to her again. "I love you, Ella," Anna said out loud. Her faith had never been stronger.

13

The Hike

The group had been instructed to take the day before the commencement of the hike to mentally and physically prepare for the arduous journey through the dense and unforgiving Sierra Nevada mountains. They had been given a list of required items for the hike, and Anna had collected most of them, but she needed to get a few last-minute items like snacks and a few extra quick-dry tank tops. Santa Marta wasn't as large or developed as Cartagena, but she was told by Hugo that she would be able to find the common items she was looking for at the local department store.

Anna was immediately struck by how different a vibe Santa Marta had from Cartagena. The city had parallels to San Jose, and most of the areas in downtown were eerily quiet in the evening. The tourist areas were the only ones that were safe at night, and Hugo was careful to show the group the streets that bordered these safe zones. As Anna was entering the Santa Marta hotel, she met a man from Nova Scotia who had told her he was pickpocketed just a block from the hotel. He told her to be very careful with her money. He was busy trying to rebuild his life after the loss of his phone, using a borrowed laptop in the lobby of the hotel, and Anna felt terrible for him.

Much of the historic center of Santa Marta was in disrepair. There were a few buildings, like the gold museum, which had been well maintained and cared for, and the harborfront was also lovely, but it wasn't a place Anna would want to spend too much time in. Walking around, even during the day, she felt as though she had a target on her back. Men would crassly shout, "Rubio" or "Princesa" to her everywhere she went as a comment on her blonde hair, and this made her feel objectified and bitter towards them. At first, she would angrily give them a nasty look, but then realized that was giving them what they wanted, her attention. Holding her head high, she eventually smiled and ignored them as if she didn't have a care in the world.

Every thirty seconds buskers would run up to her trying to sell her everything under the sun, from tours to massages. People were also trying to sell things everywhere in Cartagena, but it felt innocent, and they were not as pushy. In Santa Marta, it seemed as though if she said or did the wrong thing, it could easily boil over into something more sinister. In general, Anna wasn't a fan of the city of Santa Marta and was glad that they would only be spending a day there.

THE SMALL TOURIST street of Santa Marta was, by contrast, quite beautiful and reminded Anna of Getsemani in Cartagena. Brightly colored flags, and umbrellas were fashioned as sun canopies over the narrow historic street. Carved Spanish colonial woodwork was everywhere amongst the fine art graffiti galleries. The Teatro Santa Marta was a large Art Deco building in pink and green pastels that looked like something out of the nineteen forties, and the gelato shops that surrounded it were even better than the ones in Cartagena.

Gnam Gelato was a delightful reprieve from the chaos and incessant murmur of the street vendors. Mojito gelato with cocktail umbrellas, limes, and mini bottles of rum on top looked sumptuous, and Anna got a double scoop of it to refresh her as she continued her walk to the main street where the stores were located. Carrera Five was an absolute zoo of people, moving in all directions on foot and by bicycle. Anna easily found all of the items she needed at the department store, which felt very much like a South American version of Walmart, and decided she would spend the rest of her free time in the relative quiet of the harbor until dinner.

It always seemed to Anna that the waterfront neighborhoods of every city felt the most peaceful. Logistically this made a lot of sense, they were missing a physical side of urban life entirely. Open-ended areas of the city, invite large oceans or lakes of tranquility into them. The waterfront of Santa Marta was no exception. Pristine marinas, lively beachside eateries, and small packed beaches full of sunbathers were bordered by a very well-maintained promenade. Triangular huts selling fruit juices, sarongs, and jewelry, the area felt fresh and clean in contrast with the city itself.

Anna visited one of the huts and decided to try a Guanabana milkshake. Her guide had introduced them to the fruit the night before, and she was curious to try it. The juice was smooth and creamy with a refreshing tartness and bits of sweet grainy flesh. It was a different flavor to anything she had tried before, but she loved it and ordered another. To Anna, discovering and trying new foods had always been one of the best things about visiting another country. During her time in Colombia, she had also discovered one of her new favorites, the coconut lemonade, which was to die for, and the famous 'Arepa' which now topped her list of favorite foods.

AS A LAST SUPPER before the morning trek, Anna met Priya and Yves for dinner and ordered the largest, cheesiest margarita pizza in town, smothered in basil and tomatoes. They were going to be well-fed during The Lost City trek, but they had no idea what they would be getting. They toasted their upcoming adventure and washed their carbohydrates down with a round of local beers that were deep and malty. They were all feeling a bit nervous about the impending hike, but there was also an air of enthusiastic anticipation.

"I'm so excited about tomorrow. Do you think we're going to have to sleep in hammocks, or do you think we'll get bunk beds? I don't want to sleep in hammocks because I think the rocking and swaying would probably keep me awake, and I'm going to need all the sleep I can get to make it through this trek," Priya said.

"When I was talking to Hugo yesterday, he said that it's low season, so we're probably not going to have to sleep in the hammocks, which is good news because there will be fewer mosquitoes too. The bunk beds are equipped with special mosquito netting that Hugo said is good," Yves reassured them both. Anna had heard horror stories about the mosquitoes in the jungle and prayed that it wouldn't be too bad. She had packed a lot of repellent just in case and had some long-sleeved shirts and leggings for evenings in the camp.

BACK IN HER ROOM at the hotel, Anna went through her pack again to make sure she was bringing the absolute minimum on her hike. She had read that every gram counted when you were hiking ten-hour days. She had boots, hiking poles, a waterproof backpack cover, a water reservoir in her pack to stay hydrated, two pairs of shorts, three sets of tank tops and sports bras, a little bottle of detergent, toiletries including sunblock, a wide-brimmed hat, and five pairs of socks and underwear.

Anna had also been told that everything gets and stays soaking wet in The Lost City, either through rain or sweat, and the twenty-four-hour humidity of the jungle, so she should bring as few items of clothing as possible and be prepared to carry them in a ziplock bag from camp to camp. With everything packed up, and her water reservoir filled, her twenty-liter bag weighed over ten pounds. It felt heavy, and she hadn't even started walking, but with the poles, it was much easier to manage. She was a little concerned about the weight but was carrying the bare minimum so would just need to make do.

The off-road vehicles that would take them to the start of The Lost City trail, two hours away, were going to be picking them up at the ripe hour of five am. Anna tossed and turned, nervous about the adventure ahead and the distance that they had to cover to reach the city itself, which was over 30 km. She also knew about the elevation they would be hiking, but she had no idea what it would feel like to do it. 'Is my pack too heavy? Did I miss anything? Am I too old to be trying something like this?' Anna's monkey mind could not be silenced into the late hours, and she probably got four hours of restless, broken sleep at the very most.

The one dream Anna could remember having was of Ella, running across the park through the lush grass and stopping, panting with her pink tongue gleaming, at Anna's feet. Anna had laid down on the ground, the cool of the leaves of grass on her neck, placing Ella on her chest. They both drifted peacefully off to sleep together as they stared up into the rich blue sky, happy.

FIVE IN THE MORNING arrived too soon, and groggily, Anna grabbed her crammed day pack, poles, and carry-on bag and went upstairs for a quick breakfast of eggs, fruit, and toast. She tried to eat as much as she could physically hold, to store energy for the long trek. Stowing her larger bag in the hotel locker room, she climbed into the back of one of the jeeps and put her gear under the seat. 'Okay Anna, it's game day, let's do this. El is with you, so you can and will complete this trek.' she thought, trying to encourage herself and get ready for the day ahead.

The vehicles had arrived at the local town eatery in El Mamey, from which they would take their first step of the journey, around eight-thirty. Another hearty meal of rice, beans, and fish was waiting for them. It seemed like they had just eaten, and it was far too early for lunch, but Hugo encouraged them to eat again, "You're going to need all the calories you can pack in, trust me." Anna ate more than she could remember and wondered humorously whether all the extra weight would make the trek easier or more difficult.

AFTER LUNCH, the group stood around, hiking poles extended, packs on, and boots tightened. They listened intently to their guide, Elgin, prime them on the trek, "Day one isn't going to be easy, but if you can get through the first two hours and Devil's Nose, you're golden, and you will be able to make it the rest of the way to Ricardo camp. It's going to be hot and sticky, but the first bit is the worst. Let's get going; we have a lot of ground to cover."

They started making their way down the arid road, with no trees, and the blaring sun bearing down on them, scorching the dry dirt under their boots. 'Alright, this is nothing,' Anna thought, 'I can handle this.' Energized by the shared buzz of the group, she carried on with a steady, quick momentum, matching pace with the others.

HALF AN HOUR later, they rounded a bend, and before them stretched a mountain path winding its way straight upwards at a forty-degree, severely accented angle. "That's the infamous Devil's Nose!" Elgin said gleefully, "Here we go, onwards and upwards Pumas! You've all got it in you to make it to the top!" he continued.

"What in heaven's name is this desert nightmare?" Anna said aloud, staring up at the sharp path laid out before her.

"Awe, it isn't that bad, just take it slow and use your poles," Hugo said encouragingly, as he started moving up the path with the group.

'I don't have any other choice than to take it slow,' Anna thought. She took one small step after another, using the poles and her arms to pull herself up to take some of the weight off of her legs. 'I'm going to die on this trek on the side of the mountain as a puddle of salt and gear. I wonder how many people don't make it past day one,' she said silently to herself as she continued to move up the mountain.

The group had all disappeared ahead of her, and it took everything in her to fight the battle of her mind, squash her negative thoughts, and continue the climb. The unshaded sun beat down, as buckets of sweat poured down her face and back. "Keep following what was lost. Keep going," Anna chanted with each step. She was becoming disillusioned and wanted desperately to quit, but more desperately to keep going. She saw a small caramel-colored creature on the side of the mountain, under a dry bush.

Moving with more intention, and picking up her hiking speed, Anna saw that it was a black and tan-colored puppy with a beautiful, sweet face. The dog barked once at her and bounded up the mountain, turning to stare at her before disappearing around the bend and out of sight. "El, you must have sent her to encourage me!" she exclaimed. Finding renewed inspiration and verve, Anna pulled herself up and over the sharpest incline of the mountain.

As she finally rounded the top corner, her legs and arms were aching and burning. Anna saw the Pumas group resting at the top, rehydrating, and happily went over to join them. "Yay, Anna!" Priya said encouragingly, with a wide smile, "We made it through the worst part of today. All smooth sailing after this."

The rest of the day was indeed easy, compared to the torturous challenge of Devil's Nose, and Anna was glad that she hadn't quit. She was even more grateful for the quiet company of the black and tan-colored puppy that had followed her, from afar, all the way to Ricardo camp. Anna had a quick, rushed dinner, and fell asleep in her bunk bed at seven, before anyone else. She was entirely spent but full of gratitude for having made it through the day.

14
The Companion

Well before sunrise, a soft male voice chanting a Spanish song woke Anna from a deep slumber, and she could see the silhouette of Hugo moving around camp, his singing growing louder and louder as he went. "Buenos dias Pumas, rise and shine, it's time to face the day! We have breakfast and delicious coffee waiting for you downstairs. It's a long journey today, so get moving," he chanted. Anna suddenly felt like she was in the army, ripping the mosquito net out of the side of the mattress and rolling out clumsily onto the wood floor, landing with a thud beside the black and tan-colored puppy who had been camping there all night.

"Oh hey girl, sorry, you can keep sleeping! I didn't mean to wake you up," she said as the dog excitedly got up and started wagging her entire backside vigorously, jumping up onto Anna. 'What a wonderful way to wake up, I miss waking up this way. I know she is still with me, but do I ever miss her furry little body of energy in the mornings,' Anna thought as she brought herself to her feet and stretched. Despite it being a rather cramped bunk bed, she had managed to get a deep restful sleep brought on by the previous day's exhaustion.

Breakfast was a hearty scrambled egg, corn tortillas, bread, and fruit with juice and coffee, and they were on the trails by five thirty before the sun had made its appearance. The morning mist and lack of sunlight were cool and comfortable; the air was still thick with humidity but compared to day one it was nothing. The first hour would be a relaxing way to ease into the long eight or nine-hour journey, depending on pace.

As the sun came up and started bearing down on them again, they neared the base of the first mountain of the day. After an hour of steady climbing up an incline that varied between thirty and sixty percent, Anna was once again squarely at the back of the pack of Pumas, by herself. She placed each step intentionally on the rocky terrain and used her poles to lift herself up and forward steadily, leg muscles burning, and back aching from the weight of her pack.

HOURS WENT BY as she climbed alone, seeing only native men and children pass her on horses. They were focused and wouldn't smile at her and stared straight, deadpan expressions on their faces. Covered in sweat, exhausted, and alone, Anna reached a river that offered a clear view ahead of the path for about half a kilometer. Seeing no one else, she started to cry. 'I didn't sign up to do this all alone. If I'd wanted to make this trek alone I wouldn't have joined a group,' she thought, sorry for herself, 'Everyone in the group seemed so nice and now they've all left me in the dust. So much for us all having to act like a team and stay together.'

Anna took a shaky step onto the suspension bridge that would carry her across the river, tears in her eyes, and feeling utterly defeated and disillusioned, wanting to give up the fight. Halfway across the wooden bridge, she looked down to see she had been joined by the black and tan-colored spotted puppy with giant satellite dish ears and warm caramel eyes. She knew this beautiful girl had followed her all the way from camp, and she was truly Anna's hairy little angel.

"Hello, beautiful girl. Are we going to do this together? Do you know my beautiful Ella?" Anna stared into the empathic eyes that seemed to blink in acknowledgment, "Ok let's go do this, my new friend. Thank you for being here for me, I am going to call you 'Caramel' because you are so sweet." Caramel looked up at Anna as she trotted alongside her, matching her stride as they continued to make their way along the bridge to the other side of the roaring river.

ANNA CONTINUED the hike with Caramel, and her spirit felt renewed and energized by the quiet companionship. She no longer felt bitter towards the group for leaving her and relished the time immersed in nature with her newfound friend. She started to remember what the elder Ezquierda had said to her, 'You are the natural world and it is you. Everything is one. You have an animal spirit guide close to you. Her divine female energy is a part of you and you should listen to it carefully.'

Anna felt like she was putting the elder's guidance into practice now on The Lost City trek. She would stop periodically to close her eyes, breathe, and feel the essence of the natural world moving through and around her, coursing through her veins. She could feel her animal spirit guide, Ella, surrounding her. Tapping into the spiritual energy charged Anna up, and electrically charged chills would spread throughout her entire body, renewing her.

When Anna would start to tire and slow down again, Caramel would run up ahead on the path and look back encouragingly, waiting. It seemed that Caramel was pulling her along, and lending her strength. They soldiered on, and by noon they had reached Camp Mumake where Anna saw the rest of the group had just arrived and was taking off their packs. 'Why didn't I see them ahead of me? That's so strange,' Anna thought.

"Anna, you've always got a hairy entourage with you; it looks like you're like the dog whisperer," Yves called out to her. Looking down at Caramel, Anna smiled gratefully and thought to herself, 'No, I'm just getting back in touch with my true nature and the spiritual world. Thank you, Caramel, and thank you, dear Ella. I wouldn't have gotten here without you.'

AFTER THE GROUP finished their lunch, they quickly got ready to make the second half of the eighteen-kilometer journey.

They would need to cross the Rio Buritaca twice before arriving at Camp Paraiso, and the guides were not sure what the water levels would be like after the rains. "We can usually make the river crossings on foot, but the worst-case scenario is if the water levels are too high and the current is too rapid, there is a homemade cage that we'll pull each of you across the river in," Hugo cautioned them.

Anna was hoping that the torrential downpours over the past month hadn't rendered the rivers uncrossable. Something was unnerving about the thought of having to cross a raging river in a small homemade cage, rigged up to a single cable. This was nothing like the zip-lining adventure in Costa Rica. Who knew if the rig was even maintained? The group soldiered on, Anna at the back with her new loyal friend Caramel.

AS THE GROUP neared the first crossing at Rio Buritaca, it didn't look good. Swirling rapids were churning through large boulders, and the water level looked ominous and high. The mules that had joined their group carrying supplies would need to make the trip across the river, but it was decided that the hiking group would need to go by way of a rusty metal cage.

The cage was slung precariously on a cable, swaying in the wind, meters above the rapids. A few rotting wooden boards were the only solid material that would separate them from a swift ride down the river. Vines twisted their way down from the forest canopy into the torrent, being pulled along and down the raging waters. They waited half an hour for the safety harnesses and helmets to arrive, which the group was very grateful for.

Anna and Priya were going to be the first to cross the river in the cage, and they were fitted into a harness and helmet as Caramel looked on curiously. No words were exchanged, just nervous glances and expressions, with silent prayers. Climbing an old wooden ladder up to the platform from which she would be attached to the cage, Anna placed her feet onto the old boards and gripped the sides of the cage tightly as it bounced up and down under her weight.

She was then clipped to the top of the cage before the two men holding it in place let go, and she went ripping across the river. The cage's momentum slowed down, and she bounced backward over the rushing currents as the man who had been the first to pull himself to the other side of the river quickly caught her with another rope attached to the bottom of the cage and started pulling her back in the right direction. Anna's legs were shaking, and she was scared stiff as she reached the large rock where she would exit the cage and climb down to safety.

When she was down and free of the harness, she looked back across the river to see Caramel pacing back and forth along the bank on the other side. Anna wondered how she would get across the river and felt bad for her. She already missed her dear new friend. The mules had started to make the river crossing, laden with heavy loads of supplies. As the first mule neared the other side, a rapid caught him at the wrong angle, and he was knocked off his hooves down into the water, taking his handler with him.

ANNA SCREAMED and watched in horror as the huge animal struggled to get to its feet as the man who had been taken down also screamed and struggled. Moments later, the animal managed to find its footing on the slippery rock bed and climb onto the bank, ears straight back, soaking wet. Anna ran over to the riverbank to see if the man and the mule were alright. The man said he was fine, but Anna could tell that the mule was very shaken and she did her best to comfort her.

It took over an hour to get the group and their supplies across the river using the rickety contraption, and they were on their way again. One more river crossing, 'I hope that it's not nearly as challenging or time-consuming as this one,' Anna thought. She was sullen as they started to walk away from the riverbed, and she wished that Caramel would be joining her. Looking back across the river, Caramel was running fast in a full gallop south along the riverbank.

THE SECOND CROSSING of Rio Buritaca was easy compared to the first one, and they would be able to cross on foot. Anna removed her boots and socks and was able to clip them to her pack and make the entire crossing barefooted. It was a slow crossing aided by a human chain of guides who had strategically fanned out across the river, but in the end, everyone in the group had done it without having to use the good old rusty metal cage.

THE HIKE CONTINUED through the afternoon and took them along the side of the mountain through hay-colored hills of farmers' fields dotted with cattle and very little shade. Anna found herself alone at the back of the pack again, and she prayed for the strength to keep going as the muscles in her legs, ones that she didn't even know she had, screamed with every aching step forward.

"Ella, darling," she said deliriously with heat and dehydration, "please, baby girl, help me now. Just a few more hours. The sun is so hot, and I'm out of water," Anna said as she fell onto the ground.

She must have blacked out. A few moments later, lying on her back, a giant blinding yellow orb pounding down was replaced by a giant tongue and two beautiful caramel-colored eyes. "Oh, my dear angel, Caramel! Sweet baby girl," she said, relieved and confused. "What are you doing here, sweet girl, and how did you get across the river?" Anna said out loud to Caramel as she pulled herself up off the parched ground. She was starting to wonder why she hadn't heard from Ella in a while, but she kept walking, taking slow hot steps, Caramel by her side.

BY FIVE IN the afternoon, as the sun was hanging low in the

sky, Anna and Caramel finally reached Camp Paraiso, well behind the group who had already grabbed refreshments and sorted out their sleeping arrangements. Anna crumpled onto the bench of a picnic table, letting her pack slide off her back, hiking sticks falling to the floor. Young Caramel was still bright and full of energy, staring back at her lovingly.

"Yes, I'll get you some water, and you can be sure you're going to get a good portion of my meal too," she told her. The guides had cautioned them against feeding the animals at the camps, but Caramel wasn't just an animal; she was a loyal friend to Anna when humans had seemed to let her down again. Caramel seemed to understand what she needed.

Anna was in bed and asleep by seven, physically spent from the day's events and the journey to Camp Paraiso. Caramel was curled up under her bed cozily, snoring. The next day would be Anna's forty-third birthday, and the group would be making the final big push to The Lost City, 'La Ciudad Perdida,' where they would make the final ascent and climb up over twelve hundred ancient steps to the top. Anna dreamed of Ella and Caramel chasing each other through the dog park close to her home, barking joyfully as she laughed and smiled at them both.

FAITH   FEAR   FORTUNE

Caramel the dog, The Lost City, Columbia

15

The Rebirth

Five in the morning was punctuated once more by Hugo's wake-up song and a flick of the bright overhead lights above the rows of camp bunks. It was Anna's birthday, and the day they would enter The Lost City; she was energized with the promise of what was to come. In speaking with the local guides, they explained that the ancients believed The Lost City was sacred, and they had traveled there to leave all of their pain, heartache, and anguish. It was a spiritual pilgrimage for them. Anna felt that the spiritual world was giving her a gift by allowing her to spend her birthday in this sacred place, to hopefully release and leave all of the pain of the past year, and she was extremely grateful for the divine coincidence. She was determined not to take any of it for granted and would pray and meditate once she finally reached The Lost City.

The Pumas rose, ready to face the day, and congregated around the picnic tables at five fifteen for a delicious breakfast of eggs, tortillas, and strong Colombian coffee. "Yesterday was tough, but today is going to be no break from the intensity, team. It's going to be a hot hike to the base of The Lost City that will take us approximately two hours. From there, you're going to have to climb up over twelve hundred steps, and don't let the word 'steps' fool you; they are dilapidated two-thousand-year-old stones that have been carved out of the rock into the steep mountain face. They will be mossy and wet, and crumbling, so you're going to need to take your time. There is no race here, especially on the steps," Hugo set the stage for the ensuing challenge of the day.

ANNA WAS PUMPED to be spending her birthday in The Lost City and was happy that no one seemed to know about it. She was always shy about attention, especially on her birthday, and preferred quiet celebration over any sort of extravaganza, no matter how small. She finished her eggs and gave Caramel some large chunks of tortilla and pieces of egg, covertly under the table. "Here you go sweet girl, we need our energy today!" she whispered softly as Caramel licked her hand in gratitude.

Donning her pack and tightening up her boots for the final leg of the journey to 'La Ciudad Perdida,' Caramel by her side, and Ella in her heart, Anna took her place at the back of the Puma pack and began the hike. Birthday or not, she didn't feel older and was happy to be doing something that would challenge her ingrained notions of aging. Here she was amongst people who were at least half her age, and she was not far behind them, completing what was one of the toughest-rated hikes in the world. She felt confident and energized, ready to take on the day and any challenge that lay before her.

TWO HOURS into the hike, that indefatigable Colombian sun was burning a hole into her resolve and her positivity was waning. Anna was again far behind the back of the pack with loyal and wise Caramel by her side. As she reached The Lost City steps, a beautiful pair of native children ran past her, wearing black rain boots above their knees, with muddy white tunics and adorable mini mochila bags, their dark hair cut into bouncing bobs. "Ola," Anna managed, exhausted through sweat, but as she looked up again, they had disappeared up the stairs into the overgrowth. Caramel started bolting after them up the steps and disappeared into the thick jungle after them.

"No, Caramel, wait for me," Anna yelled, legs heavy and aching as she pushed her way forward up the first section of steps, maybe fifty or so. Caramel was gone. Just when she needed her the most. Her pack was laden with humidity, heavy and cutting into her shoulders, causing them to ache. The bottoms of her feet felt like they had been pummeled with a tire iron, and the steep winding staircase that lay before her seemed to carry on in a never-ending ascension. She kept climbing, putting her hiking poles on each next step to pull her up when all of a sudden, she slipped down five steps, her hip banging hard into the side of the ancient steps.

"On my birthday, this is how it ends, God! This is how you

decided to bookend the hell that I've been living for the past year! Alone, exhausted and drained of all my physical and mental strength in the middle of nowhere in the darn Sierra Nevada mountains!" Anna let go of a primal scream at the gray, wet wall of twelve hundred never-ending steps. "I can't take this anymore, I'm even alone and outmatched on my birthday, of all days! This isn't fair. My life is a train wreck!"

She started to cry, but not just cry; Anna lay bawling in a heap. She lay beside the steep steps just above a small tree to keep her from plummeting down the rest of the staircase. Anna rocked back and forth, hugging herself in resignation. "I can't do this. It's over. Why did I decide to take on something so foolishly challenging? I don't know anyone here, no one gives a shit about me, and I could just die here on this forsaken mountain," she said to no one.

"God has not forsaken you, Anna, not even close," it was Ella's voice booming down from above her louder than ever before. "Do you know why you're here, Anna, what brought you to this place?" she continued.

"No, I don't. I thought I could do this. I thought I could do something big and prove to myself that I'm still good enough for something, and that I can make something positive happen in my life. Ella, please tell me why I'm even here." Anna cried.

"Anna, you're here to remember who you are. There once was a roaring fire that raged within you, a creative inferno, and for you now, it is only a spark. That doesn't matter, the spark is still alive there deep within your core. It's been doused by people who wanted to steal it, it's been stepped on, and sand has been thrown on the fire, but that tiny spark is still here. That's what we're here to focus on. Keep going Anna and follow what was lost," Anna looked up and saw the voice was Caramel's.

"That's right Anna, remember when the elder Ezqierda told you that you're a part of the natural world, and it's all one. Well, he was being literal. I am your Ella," Caramel said, "It's no coincidence that I'm only two months old, Anna. My spirit started merging back into this body the very day my old one gave up on me. It's a process. In the meantime, I was still able to talk to you through the Existent Spirit." Anna was dumbfounded and couldn't believe she was hearing. Caramel was speaking. Ella was speaking. Ella was Caramel.

"What? Did this mean that Ella never really died? You never really died?" Anna cried in tears of joy.

"No, of course not, Anna. I could never, or would never leave you. This body certainly isn't as well-groomed as my last one but Anna, yes, it's me. It's horribly cliche, but love never dies, especially the love between a canine spirit and a loving human one," Ella said.

"That's why your eyes were so familiar! Why am I so blessed, why is this happening to me of all people? Thank You God for this miracle," Anna was now crying tears of pure joy, what fortune had life delivered her.

"It happens all the time, Anna, but do you think if people talked about it, anyone would take them seriously? Would you tell anyone this is happening? You can sit all day on the side of this mountain or we can climb up to the top together, like in old times. My legs are young and spry now; you wouldn't believe how fast I can fetch a tennis ball, faster than ever before," Ella replied with a twinkle in her caramel-colored eyes.

ANNA STEADIED HERSELF on her feet, holding onto the small tree on the side of the steps, tears of joy running down her cheeks, "Ella, I love you so much. Let's go, my little bundle of delight. Thank you dear God for this gift, thank you."

She looked back down at the fifty steps she'd ascended and saw Hugo, who could see that she had been crying. "Hugo, do you mind taking my hiking poles for me? I fell while using them, and I'm afraid to continue with them," she asked. "Of course, Anna, it is your birthday after all," he smiled and winked back at her.

"How on earth do you know it's my birthday?" Anna laughed. "It's on your application when you sign up for the tour. I'm really happy to see you have company on your special day," Hugo replied, "And I'm right here behind you, Anna. You are not alone."

"Thanks, Hugo, and yeah, we're old friends. Very old friends," Anna said as she steeled herself and placed her foot on the next step, Ella by her side. Using her hands to steady herself, Anna half-walked half-crawled up the twelve hundred steps, but it didn't matter, she was doing it. With every slick step she kept from sliding down, she was closer to something she never thought she would be able to accomplish, and on top of it all, she had her best friend back.

Standing at the very end of The Lost City staircase, Ella beamed at her, butt wagging and tongue hanging out in glee. "Come on Anna, you're almost there. Just a few more steps," she could hear Hugo calling.

AS ANNA REACHED the top of the staircase, she felt the weight of the year of pain slide down off her back, and she was surrounded by an intense white light. The light was illuminating the entire Lost City and gave it an ethereal glow. The stepped terraces radiated up in circular patterns, and as she stepped onto them, she felt herself grow weightless as her soul seemed to sail up and away from her body, carried on the breeze.

The faded impression of the elder Ezquierda came to her, and at once, Anna understood what it had all been for. She felt the entire universe expressing itself through her and understood that the past didn't exist and held no power over anyone. All negativity and pain were rooted there in the past, which was an illusion. The only thing that mattered and would ever matter is our unified spiritual love for one another in the present moment. The divine energy of the elder Ezquierda spoke to her without words, and from a place of absolute love and knowledge as she felt the warm ball of Ella's radiant energy join her heart.

"Elder, can I ask you a question?" Anna projected.

"Yes, Anna, of course," the elder energy returned.

She asked the elder Ezquierda one thing, "Elder, is there anything you can do to bring me human love? I'm eternally grateful for Ella, I don't even want to ask this because it will make me seem ungrateful, but I've just been lonely for a very long time and I feel restless."

"You know that thing you have where you can communicate with animals without using words? Well, that is called tapping into the Existent Spirit energy that binds us all together in unity, and you can use that to manifest your destiny as well. It is what drew Ella back to you. If you want something to be, it will be, Anna. Awareness is the only way to bring about change. You need to know, however, that because you are part of one, you are already whole," the elder replied.

As Anna drifted back into her body, standing on the top terrace of The Lost City, she felt the past, and the weight of her pains sink into the soils of the city and join with the pains of generations of others who had made a pilgrimage to this sacred place to cast aside their darkness. Anna was free. She gazed at the hazy, misty mountains and felt only deep gratitude for her life.

AS SHE DESCENDED back down the terraces, she caught sight of the Pumas standing together, taking photos and ran to join them, light as a feather. "Anna, I have something very important to tell you. Pumas, you know what I'm talking about," Hugo said excitedly with a giant smile on his face, as the entire group lit up and joined him in singing Happy Birthday to Anna.

"Anna, I hope this is one of the most memorable birthdays you could have, in a beautiful place. May this year bring you fortune and happiness. Pray and set your intentions for the year in this sacred place, and they will come true," Hugo added after the group had finished singing.

"Thank you everyone! Yes, this has already been the most memorable and indeed fortunate years of my life, and it's only day one. I have only gratitude for you all, and all of the blessings in my life," Anna thanked them through tears of elation, "Oh, and everyone I would like you to meet Ella. Her name was Caramel but she told me she is more of an 'Ella' than a 'Caramel.'

Ella, her bright caramel-eyed companion for eternity, jumped for joy as Yves grabbed her face and gave her a huge embrace. Ella made the rounds, being sure to get as many pets and ear scratches as she could, and gave plenty of kisses in return.

AS THE GROUP finished their visit to The Lost City and started making their way back to Camp Mumake, Anna felt only grace and peace in her body, heart, and mind. Ella at her side, they made the trek with the group and arrived at camp just before the torrential rains started. After dinner, Anna suddenly caught sight of a mysterious box that Hugo grabbed on their way to The Lost City days earlier. He rushed with it to the kitchen and moments later emerged with a birthday cake, and the entire dining hall engaged in another round of happy birthday as Anna's cheeks grew red. "How did you get this cake of all things up here? This is incredible, thank you so much! I can't believe you managed that," she said to the group.

As she stared at the flame and was told to make a wish, Anna was reminded of that one spark deep within her that she had started fanning back into a roaring flame. She closed her eyes and concentrated deeply on the Existent Spirit energy that the elder had put her directly in touch with. She manifested her intentions for the year as she blew out the candle amidst cheers and claps, feeling deeply that her life story had turned for the better. She could feel the Existent Spirit energy coursing through her, and her mind was clear except for a deep gratitude for her life, and for Ella having found her way back to her again.

Part 3: Fortune

When we've found our faith, and
faced our fears, we will find our fortune

The hike back from The Lost City ended with a well-earned celebratory lunch and many refreshingly crisp rounds of Colombian beer. Never had a beverage tasted so revitalizing to any of them, and the Pumas cheered and toasted the monumental achievement of completing the trek. The total journey had been an arduous one, and Anna had clocked it at ninety total formidable kilometers. It might have been torture, had it not been for her dear companion, Ella, rejoining her in the depths of her journey.

The new, younger but larger version of Ella would eventually be joining Anna on her return trip home to Toronto, but first, they would both be taking on a new adventure together in Argentina. They said their goodbyes to the members of the Pumas hiking team, as well as their dear guide, Hugo. Anna was ready to strike back out into the unknown with her treasured companion by her side. She decided to make the journey back to Cartagena by bus from Santa Marta, with Ella curled in her lap. The four-hour journey would give them both time to rest and reflect.

Before she had embarked on The Lost City tour, Anna had been worried about having to go home and face the emptiness of her Toronto condo. She had booked one more hiking tour that would take her through the mountains of Patagonia to cap off the trip. She had always longed to see glaciers, and as she had inched further and further south on her travels, Patagonia had made more and more sense. At the time, Anna felt like she didn't have anything to go back home to, although this had fortunately changed.

BACK IN CARTAGENA, Anna took Ella to a local veterinarian for her required paperwork, microchipping, and vaccinations. The vet would be registering Ella as Anna's emotional support animal so she would be able to travel in the airplane cabins with her. Ella's new body was going to be considerably larger than her miniature schnauzer one, and Anna hoped this wouldn't cause problems in the airport. Ella was still just a few months old and already twenty-five pounds.

Ella seemed to love having her young joints and cataract-free eyes back and had enjoyed playing fetch for at least two hours a day with Anna. She was grateful for every single second and had missed these park dates immensely. It was as though no time had passed, and Ella was faster than she had ever been before at retrieving the ball. It had been over ten years since they had been able to play so energetically like this together, long before Ella had been spirited back into her new body.

They had spent two weeks at a home share rental in Cartagena when Ella's paperwork finally came through, and they were able to book the flight to Buenos Aires, connecting through Panama City. It was still going to be over a week until they would be joining the tour to Patagonia, so there was plenty of time to explore Buenos Aires together. Anna had heard only wonderful things about the city from former colleagues and acquaintances. She had purchased a beautiful new red leather collar and leash for Ella with a gold-engraved nameplate that said "Ella" on the front and "Also known as Caramel" on the back in small letters, as a joke. Ella displayed it proudly at the airport as they checked in and boarded their flight. Ella curled up and fell asleep on Anna's lap as the plane departed for Buenos Aires.

During the seven-hour flight, Ella was the star of the show. The stewards brought her a small dish of water and some jerky, and every child on the airplane came by to visit her and give her a scratch behind the ears. "What a beautiful puppy, and she is so well-behaved! How on earth did you get her potty trained for the flight so quickly," a steward had exclaimed.

"Oh, she is an old soul," Anna had replied, in total honesty. Flying over Panama, they marveled at the endless greenery with rivers snaking their way through the densely forested land, specked with small fishing boats. Large freighters were fanned across the entrance to the world-famous Panama Canal. When they finally reached Buenos Aires late in the evening, they caught a shuttle from the airport to their apartment. Anna had booked a small loft close to the heart of downtown, which had seemed like a good base from which to explore the city. It had been a full day of travel, so after letting Ella out to relieve herself, they promptly went to sleep.

BUENOS AIRES was a bustling metropolis full of vibrant people, music, dance, and art, rooted deep in its rich Spanish colonial history. It was founded in fifteen thirty-six when Spanish sailors arrived in the Río de la Plata, but came under attack by indigenous people and was abandoned only five years later. The Spanish had sailed into the port of Santa Maria, it was later founded again in fifteen eighty by Juan de Garay. A city built on rebirth, Anna could feel a kinship with the place. The houses along El Caminito were painted in vibrant primary shades of all colors, saturated against an azure sky, a stark contrast to the dull gray skyscrapers at home that set the backdrop of winter in Toronto.

Anna had heard that the houses on El Caminito were painted this way originally because the poor were forced to use the lower-cost, brightly colored leftover paint. Over the years, by happy accident, the bright street became a tourist hotspot, and the tradition had endured. Another example, Anna thought, of how the root of struggle can become something reclaimed and beautiful. The spirited homes of the area seemed to evoke the playful imagination of some of her favorite childhood authors, like Dr. Seuss.

Ella met hundreds of other dogs as they explored the city's seemingly never-ending spaces of emerald green. Buenos Aires had over two hundred sprawling urban parks to choose from. Sitting on empty public benches amid deeply saturated, fragrant blooms and chatting with locals while enjoying coffee and a dulce de leche cookie sandwich became a daily ritual for Anna and Ella. Admirers of strong female icons like Eva Peron and Frida Kahlo, the portenos of Buenos

Aires also seemed to be long-standing defenders of equal rights for women, and Anna felt both physically and psychologically safe and comfortable in Buenos Aires.

Anna could somehow feel the Existent Spirit energy everywhere in this place, a current of light running just beneath the surface of the physical dimension they could touch, see, and taste. As she went to sleep each night, caressing Ella, she thanked the Existent Spirit energy and the God that manifested only happiness for them in the present and future. The past no longer had its jaw-like, suffocating grip on Anna; this had all been left buried with thousands of generations of human pain and suffering in The Lost City.

AFTER THEIR WEEK of exploring the city together, it was time to meet the Patagonia hiking group in the lobby of their hotel. Anna's normal anxiety at meeting brand-new groups of people had almost completely vanished because she had Ella's calming companionship by her side. The negative narratives and self-defeating internal dialogue were buried in the past, along with a part of the old Anna. It was a new group of eight including Anna who would be making the trek together from Buenos Aires to Los Glaciares National Park. They would be taking a small ten-seater Gulfstream airplane to the town of El Calafate in the Argentine province of Santa Cruz, and from there, they would be making a few multi-day hikes from between ten and thirty kilometers per day. Anna felt the ninety kilometers she had clocked in The Lost City was ample training for the impending mountain hikes.

It was going to be an entirely different climate and hiking experience from Colombia's Lost City. They would be transitioning from lush green humid jungles to cold icy blue glaciers and rugged mountains. Opaque humidity hanging in the tops of the trees would be traded for opaque icebergs bobbing in chilly lakes. The temperature would be just as marked a departure, from forty-five degrees to below zero degrees. This had made packing a carry-on an even taller order, and Anna was grateful for packable down jackets.

Waiting in the hotel lobby, early as usual, Anna and Ella started to see the new crew of hikers meandering into the lobby. There was a married couple from New York, Robert and Mathew, a man from Denmark named Morton, two female solo travelers from Spain and Germany, Sonja and Henrietta, a designer named Ben, and then there was Richard, from London. Anna noticed Richard the moment he walked into the room. It was as though at once the air had been sucked out of the room and it had become a static vacuum. He was tall with a slim build, sandy blonde hair, and piercing green eyes. She was immediately struck by the feeling that they had met somewhere else, maybe during past travels, perhaps a very long time ago, and she just couldn't remember.

The tour leader's name was Juan, and after going over a quick rundown of the itinerary for the Patagonia hiking tour, it was time to make introductions. 'Hurry up, I just want to hear about Richard!', Anna kept thinking as the introductions went around in a circle, 'Geeze. Finally, Henrietta! We didn't need your entire life story. You're lovely, but there is plenty of time to get to know you.'

"Hey everyone, my name is Richard, and I'm a children's author from London, England. I'm no bestseller, but I'm trying, and that's what counts. I've been hiking around Central and South America while I write my latest novel about a family traveling through Mexico in their airstream trailer. It's helping me get inspired! Happy to meet you all," Richard introduced himself.

"Hi, I'm Anna, from Toronto, Canada. Funny coincidence, I've also been hiking through Central and South America and just came from Colombia. I wonder if we have been to any of the same places Richard? Very excited to meet everyone and to be hiking the mountains of Patagonia with you. This is my dog, Ella. She is a seasoned hiker, and we met doing The Lost City hike," Anna was the last to make her introductions. 'Oh, something in common, that will make it easier if we get a chance to talk later,' she thought, her face flushing red as she caught Richard's eye for just a fleeting moment.

Juan had chosen a tapas restaurant for their group dinner. Ella was sleeping under the table nestled between Anna's legs, poised for falling scraps. Anna was next to Juan and Sonja, at the very end of the long red oak harvest table. "So Juan, how long have you been running hiking tours to Patagonia?" Anna inquired.

"I've been doing this for almost twenty years; it's my dream job! I've been hiking with my dad since I was a little kid, and if I can introduce people from around the world to the beauty of my homeland and give them a new appreciation of Argentina, I am a happy man," Juan beamed, "It's a very rare beauty Anna, to see the glaciers up close like we will. You will be amazed, I'm sure of it. So you said you've been through Central and South America, where have you been?"

"I've been to Guatemala, Belize, Mexico, Costa Rica, Colombia, and now my final destination is Patagonia! I'm so excited to see the magnificent glaciers with my own eyes finally rather than just on the Discovery Channel. I've been dreaming of what it's like for so many years and it's always been so far away; it has seemed as though it's in another world at the tip of South America, but not anymore!" Anna replied and asked Sonja, "How about you? Have you been anywhere before Argentina on this trip? Any plans afterward?"

Sonja replied, "So I'm originally from Sweden, but I moved to Spain five years ago, and I've been mostly traveling through southern countries in Europe, like Portugal, Italy, and Greece, and naturally Spain is where I spend most long weekends. This is my first trip to South America. I've been all over North America and visited your home country of Canada as a matter of fact!"

In a slight tone of pity, Anna replied, "If you visited any time between November and February, I'm sorry for the copious amounts of defrosting and piles of layers you needed to wear. What did you think of Canada, other than the weather, that is?"

"I loved it! My favorite spot was Montreal; I was there for the jazz festival, and it was summer. I couldn't believe how hot it was there; everyone was warning me about the weather in Canada, but it was incredible, just like the south of France. I also got the chance to visit Ottawa and Toronto, for just a short time. I loved the Toronto waterfront!" Sonja explained.

"Hey, that's where I live, on the Toronto Waterfront. I've never heard Montreal be compared to the south of France, other than for the reason of their French mother tongue. You're making me look at it from a different angle. Glad to hear that you enjoyed yourself and that you made the trip in the summer. I mean, winter isn't terrible if you know how to dress for it, and you pick the right activities. There aren't a lot of countries with such an epic skiing season, and you've never seen anything as beautiful as the first snow in the maple sugar bush if you're into that sorta thing," Anna exclaimed, feeling warmly patriotic with a touch of homesickness.

Juan interjected in his thick Spanish accent to share his aspirations, "Your country sounds beautiful Anna. Imagine being able to enjoy a snowfall, in real life. I'd love to visit Toronto someday, I've been as far north as Miami but that's it. I loved Miami though, the Cuban music scene was great."

"That doesn't qualify as 'north,' Juan, even though it is technically in North America," Anna launched back, sending the rest of the table into laughter, "But before Buenos Aires, Colombia was as technically 'south' in South America as I'd ever been, so I'm not one to mock you."

AFTER DINNER, the group assembled on the street in preparation for the ten-minute walk back to their hotel. Anna was sure to hang out strategically at the back of the pack with Ella because she wanted a chance to chat with Richard. "Hey, Anna, right?" Richard was walking diagonally in front of her and took a few shorter-than-usual strides to match hers, "Your little travel buddy is arguably the cutest explorer I've ever met in my life. What's her name again? How old is she?" Richard said in his beautifully smooth British accent.

"Her name is Ella; she's still just a puppy. She is only a few months old actually. I found her when I was hiking in the Sierra Nevada mountains a few weeks ago and she just stuck to me like glue. Even when I lost the rest of the hiking group for four hours at a time, I looked down and she was always still just there by my side the entire time. That's kinda when I figured it out, come to think about it, actually," Anna said, reminiscing about the day when she started to understand the divine connection between her and Ella, aka. Caramel.

"Figured out what, Anna?" Richard interrupted her with keen interest.

"Oh, that's when I figured out that she was meant for me and I was meant for her, and that I just had to adopt her," Anna explained away her comment. 'There's no way I trust someone that I just met enough to tell the crazy story of how Ella came back to me. He would think I'm totally insane and run for sure,' she thought.

"That's an incredible story Anna; what a memorable beginning the two of you had! We had dogs growing up, and I've always wanted one, but I just travel so much that I didn't think it would be fair to have one on my own. It's great that you're able to bring her along with you. Someday I will have my own dog; they are just so wonderful, aren't you sweet Ella!" Richard said, bending down and grabbing Ella's face to give her a nice big scratch behind her ears.

"She loves scratches behind her ears. How did you know? So you were backpacking across Central and South America you said! Where did you go before Buenos Aires? What made you want to come hike Patagonia? Are you a big hiker?" Anna wanted to keep the conversation rolling along as long as possible on the way back to the hotel. She was eager to find out as much as she could about Richard. 'He is an animal person, and Ella seems to approve, that's a very good start,' she thought to herself.

"Whoa, that's a lot of questions for a brain full of beer and carbohydrates. So, yeah I've been to Costa Rica, and Nicaragua so far, and yes obviously you've gotta be a bit of a diehard hiker to sign up for something as extreme as this. Thirty kilometers of mountain range in a day is no joke," Richard came back playfully but succeeded in making Anna a tad apprehensive.

"I just came from Costa Rica, before that, I was in Colombia, so we need to trade stories. Do you think the glacier hikes are as difficult as the ones in The Lost City, in terms of difficulty I mean?" Anna said, trying her best to sound as confident as possible.

"I haven't done The Lost City, but I have done Kilimanjaro. That was tough as heck, mainly because of the elevation and the thin air. We even spread the hike out over eight days, and I still found it quite tough. I think the glaciers will offer their own unique set of challenges, although I'm not sure what those are yet. I sure am looking forward to finding out," Richard added excitedly. "I've been wanting to do this for years and just never had the time until now."

"Me too! I've been dreaming about this since I was a kid. Ever since I saw them on the Discovery Channel when I was maybe eleven years old. It's going to be an epic adventure, I'm sure," Anna joined in his excitement, smiling and trying to make eye contact for longer than a second without looking like an awkward weirdo.

"So, I'm working remotely on and off on my children's novels. How do you have so much time off to do all of this traveling? You said you've been off for four months already?" Richard inquired.

"Well, it's a long story but I'm on a break from work. If we become friends maybe I'll tell you about it. Maybe," Anna skirted around his question playfully. 'No use dragging him into the details of my past just yet,' she thought as they arrived at the hotel, and she was saved by Ella dragging her over to the grass patch beside the street. 'Ella, how did you know I needed an exit strategy?' Anna thought as she followed her.

"Ok, goodnight Anna, see you in the morning!" Richard called as he entered the lobby with the rest of the group, leaving Anna and Ella to themselves on the street. They were illuminated by the hazy blue glow of the small neon sign opposite the austere, yet staunchly plain hotel.

"Oh dear Ella. What's that phrase, 'Be careful what you wish for?' I hope I don't get too caught up in another heart-wrenching disappointment. I've had one too many of those on this journey that's supposed to be helping me heal. Don't let me do anything stupid; we both know you're my one true love," Anna cooed at Ella as she gave her a long hug before walking her back inside for the evening.

17

The Ascent

Dueling songbirds with the most unique melodies came wafting along on the breeze through the open window above Anna's bed at five in the morning. She looked down at her feet, which were still asleep, to see Ella was still blissfully in dreamland. Snoring like a freight train, she was sprawled out across Anna's legs, cutting off the circulation in her feet. Carefully sliding herself out from under Ella, Anna got up to get her phone and look through the day's itinerary.

They had an early morning private flight at seven to El Chalten, where they would get their first glimpse of the Andean mountain range. The world's longest-running mountain chain, running as far north as her last destination, Colombia, and as far south as where they were headed. Anna had read that the summit of Mount Chimborazo, in the northern portion of the Andes mountain range, was also the tallest point on Earth when measured from the Earth's core. She'd been to the northern range and now she couldn't wait to get a look at this magnificent southern range of the Andes.

Anna got some last-minute packing done in her smaller knapsack, for the relatively short eight-day trip to the mountains, before gently rousing Ella. "Hey sweet girl, it's time to get up and get some breakfast, we have a big day of adventure lying ahead of us!" she whispered to her. Ella yawned widely showing all of her gleaming white teeth, and snorted loudly. "Yes, I know, darling. I'm tired too, lazy bones." Anna continued. She opened the door of her hotel room, and Ella clumsily jumped off the bed and followed her out.

Half of the group were already having breakfast. Robert and Mathew from New York, and Morten from Denmark were sitting under the rooftop arbor covered in dense ivy, enjoying scrambled eggs and fresh fruit. "It's the dynamic duo! Good morning, Anna and Ella. The hotel waitress, who is also the wife of the manager, is taking orders over there and she can probably bring some bowls for Ella's food and water too," Morten greeted them, motioning over towards the kitchen passthrough.

"Thank you, Morten, I appreciate it. Did everyone get a good night's sleep?" Anna asked as she put her bag down on the table and smiled at Ella, who was settled comfortably under the warm hand of Robert, happily giving her affection. "Matt and I crashed as soon as we got in last night because our flight from New York was delayed by six hours out of LaGuardia the night before we got here. It was great to catch up on some rest," Robert said as he smiled at Ella.

"I could sleep another few hours but I'll be ok. I'm used to being exhausted, as a physician in the emergency ward. Yes, I'm feeling the Danish jetlag creeping in," Morten added.

"A doctor! You're a good man to have around on a trip like this. I guess we can all rest a little bit easier knowing that if any of us fall down a mountain and break an ankle we've got you in our back pocket," Matt piped in with a laugh.

"Well, I specialize in acute trauma, but I've got a whole team back in Denmark, so don't go taking any big risks, young man," Morten said with authority. With his mannerisms and maturity, Anna thought that he looked to be in his mid to late fifties. The fit gay couple from New York were each no older than thirty-five. Anna thought they looked like fitness models and hoped she could get a few health tips from them later on during the hike. They both seemed like absolute sweethearts, especially Robert.

As Anna received her breakfast of fried eggs, multigrain toast, and slices of ripe salted avocado, the rest of the group started to file in and join them at the table. "Good day, Anna! Are you well rested and excited for the day?" Richard called to her first from the other end of the table. "Good day to you, Richard! Yes, we both got a great night's sleep and are super excited for the private charter over the Andes, what an awesome experience it's going to be! I can't wait, I don't even mind getting up at five in the morning!" Anna replied, more excited than she had been with the guys sitting beside her, hoping they wouldn't catch on.

"Oh girl, you seem happy to see someone," Robert said slyly. 'Too late,' Anna thought, mildly embarrassed, as she raised her eyebrows and gave Robert an awkward smile, "No, I'm just excited for today, that's all."

"Uh-huh. I've seen that a few hundred times when Matt was trying to date me. You might be able to fool everyone else, but you don't fool me for one minute miss," he shot back at her confidently and quietly with a cheeky edge to his voice, "I don't blame you; he is pretty cute, that accent is to die for, and you guys look about the same age. I'd never ask a woman her age though."

"Forty-three, for about two weeks now," Anna offered up, pretty proudly now given all she had accomplished over the last month.

"Darling, I wouldn't have put you a day over thirty," Robert said with a confidence-boosting shocked tone in his voice.

"Thank you, I like you, we can be friends," Anna laughed happily at his misjudgment of her age, wondering how old that made Richard. 'Is Richard too young for me too? I don't need that again,' she thought, worried as she smiled at him from the other end of the table, emboldened by Robert's morning injection of confidence.

"OK, guys, I'm going to go and grab my things and take Ella for a quick walk, and we will meet you down in the lobby in fifteen minutes or so. See you soon!" Anna said as she brought her empty plate back to the kitchen, Ella springing to her feet and bouncing after Anna, fully awake after a big bowl full of fish, eggs, and fresh vegetables.

"See you soon, ladies!" Robert called cheerfully.

THE AIRPORT was bustling even at six-thirty in the morning, and Anna was grateful that they were taking a private plane and had been driven right to the runway on their shuttle bus. They were able to avoid the usual lineups and headaches associated with flying on large commercial airlines. This was a brand new experience for Anna, taking a private plane, and she assumed this was one of the reasons the tour had commanded such a hefty price tag in comparison to her other tours.

The small blue and white plane looked bright and cheerful and was dwarfed by all of the mechanical equipment surrounding it on the tarmac. They had been told that they only had the use of overhead space on the plane, due to weight and balance restrictions on the small craft, and had been instructed to pack as light as possible. Much of the group had lamented this, but Anna was already pretty used to minimalist travel and didn't mind. She did have over twenty extra hairy pounds with her after all, and Ella was getting rounder by the day and growing into her oversized puppy paws.

Anna and Ella had been given the exit row so Ella could lie down more comfortably on the floor below Anna if she wanted, but as usual, she insisted on sitting on Anna's lap. The plane only had eight rows of two, fitting a max of sixteen passengers. It felt very cozy once they were all loaded into their seats with bags safely overhead. The tiny plane, which seemed featherlight as it bounced back and forth as it taxied over to the wide runway, felt like a tin can to Anna.

FAITH   FEAR   FORTUNE

The pilot eased into the throttle for a few seconds before going all out, causing the small plane to rumble at top pitch as it gently lifted into the air, allowing the air to escape Anna's lungs as she breathed out in silent relief, although too soon. As they ascended into the clouds and the atmospheric pressure offered up pockets of air, the plane met turbulence and bounced up and down, causing Sonja to audibly shriek from a few rows back. Anna just held her breath again, holding Ella tightly until they pierced through the cloud cover and the plane smoothed out at cruising altitude.

"Ella is comfortable in airplanes for being so young," Richard said from the seat diagonally behind them. He was right; she had been in many planes before in her past incarnation. Anna had taken her kayaking when she was only six weeks old, followed by trains, buses, the basket of her bike, and finally airplanes. Ella had been a very well-traveled Miniature Schnauzer, and as a black and tan spotted mutt, her life would be no different.

"Yeah, she is a natural at traveling. Perhaps it was all of the hiking she had done as a puppy in the Sierra Nevada mountains. She probably hasn't stopped moving since she was born," Anna said, thinking that she was probably looking for Anna for weeks on the trails.

Anna hoped to understand the Existent Spirit magic that helped Ella find the right physical incarnation at the right time. 'Could it have something to do with Ella not speaking audibly to me anymore after she found me in her physical body?' Anna thought. There was something intangible, magical about The Lost City; she could physically feel the energy around her as a sensation on her skin. "Tap into the Existent Spirit energy and manifest your destiny," the elder had said. Anna wondered if she was somehow now creating the life she wanted, just by wishing it so.

Anna had always felt that she could communicate psychically with Ella when she had been a Miniature Schnauzer, just by looking into her eyes, but she hadn't heard her voice until after she had passed. 'Does this communication without words have anything to do with manifestation?' she thought as she stared out the plane window at the rolling mountains of white clouds, her arms wrapped around Ella. As the plane started its descent, Anna had her first glimpse of the Andes mountain range. It was indeed spectacular. Mountains crested with white peaks stood stately around the small town of El Calafate. As they landed on the tarmac, and taxied to the gate, the surrounding looming mountains gave the tiny airport a cozy feeling.

DOWNTOWN EL CALAFATE reminded Anna of the small ski town of Banff in Canada with the brightly painted walls and warm cedar wood buildings surrounding cobblestone streets. It had its unique flavor, of course, built around Laguna Nimez, but she was amazed that somewhere so far away from her home country could have such a similar feel. They would be staying a night before taking a bus to El Chalten, a three-hour drive away, where they would be able to kayak along the Rio de la Vueltas beside the glaciers and hopefully get a glimpse of the Fitz Roy mountains.

They checked into their charming hotel, along the main strip of the street, and dropped their backpacks off before heading further into the town to grab some lunch. Anna also needed to get some food for Ella from the grocery store, and on advice from their guide Juan, a few extra thermal layers to keep her warm during the trek. Their first official hiking day would be roughly twenty-four kilometers and it was going to be a much different style of hiking than she was used to. The weather was promised to change on a dime and they would need to be prepared for everything from rain, to snow, to bitterly cold driving wind, and sunny warm temperatures.

Anna had a down jacket, a light rain shell, and some zip-up thermals, she just needed a few more base layers and leggings to put under her waterproof pants. Never having hiked in these conditions before, she wanted to be as prepared as possible physically, and this would also help prepare her mentally for the challenge. Her trusty hiking sticks and boots would get to experience another climate entirely, and she was sure that like her, they were up to the challenge ahead.

After finding all of her supplies, exhausted from the travel day, and mindful about conserving some energy for the impending hike, Anna decided to head back to the hotel with some take-away. She decided that she was going to skip the group dinner in favor of some extra sleep. 'Maybe Richard will miss me at dinner,' Anna thought, hoping to herself that he would notice her absence as she drifted off to sleep.

18

The Descent

Anna's hotel room had grown refreshingly cool over the night, and unassisted by her alarm, well-rested and energized, she sprang out of bed at six. She got herself and Ella packed and ready to check out of the hotel and get started on the trip to El Chalten. Moving her things into the lobby, she saw that most of the group members were not there yet, presumably either still sleeping or getting ready. "We missed you last night, Anna," Juan greeted her as she passed him, on her way to take Ella out for her morning stroll.

"Buenos dias, Juan! Sorry, I missed dinner last night. I was exhausted and wanted to catch up on sleep to be in top shape for the adventure we have ahead of us," Anna explained.

"Oh, that's quite alright, Anna. We didn't have a late night anyway. It seemed that everyone was feeling pretty tired from the early morning flight yesterday and had the same idea as you. Much better to be rested so you can fully enjoy the beauty of Patagonia. You're going to have a wonderful time," Juan reassured her that she didn't miss much the night before.

The bus pulled up to the front of the hotel half an hour later, and Anna and Ella were ready to board, while the others were still finishing checking out of their rooms. She grabbed the front seat so Ella would have more space, and then threw her knapsack in the overhead rack. Ten minutes later they were on the road to El Chalten, ready to meet the beautiful winding mountain passes. There wasn't much chatter on the bus, as they made their journey through the majestic landscape, everyone was glued to the windows taking it all in. Viedma Lake appeared to be a deep blue-green against the dark rocky mountainous terrain surrounding it, the only color in view other than the rich blue of the sky; Anna found it deeply saturated and captivating.

ARRIVING IN EL CHALTEN, the gateway to Los Glaciares National Park, Anna noticed that it was much smaller and fairly quaint in comparison with El Calafate. It was going to serve as their home base for exploring the Perito Moreno Glacier, kayaking, hiking, and any other activities they might try. The group was told that they wouldn't be spending much time in town, other than to eat and sleep. The town looked interesting to Anna though, and she was going to set aside some time to go out and explore with Ella before they set out for the kayaking trip after lunch. 'Perhaps I will see if Richard wants to join us for a walk around town,' Anna thought it might be a good way to continue getting to know him.

As they filed off the bus and were waiting for their room keys to be handed out by Juan, Anna used the opportunity to ask Richard if he would be interested in joining them on a walk around town and getting some lunch, "Hi Richard, how was the bus ride for you? Wasn't the scenery just breathtaking?" she said nervously.

"Yes, it was pretty incredible. I've never seen wilderness quite like it before; it's rugged and feels like another planet entirely," he agreed.

"Ella and I are going to drop our stuff off in the room, and then I thought we could explore the town a little bit before the kayaking tour. We're also going to grab something to eat. Would you be interested in joining us?" Anna said, noticing Robert eavesdropping on her conversation with a sly smile on his face. She hoped that Richard wouldn't notice Robert's facial expression. 'But he must have realized that I'm somewhat interested in him by now,' she thought better of being concerned.

"Sure, yes that sounds great. I also wanted to wander around the town for a bit without a plan and explore. How about we meet back in the lobby in thirty minutes? Is that enough time for you?" Richard seemed to love the idea. Anna didn't really need thirty minutes, but she happily agreed anyway. 'At least this will give me time to fix my hair a bit, throw on some fresh clothing and get a bit more presentable,' she thought.

"Yes, that sounds perfect, see you then," she acquiesced while already mentally preplanning what she would change into. She hadn't brought very many nice outfits, as she was anticipating spending the trip alone. Anna somehow didn't think Richard would be one to judge her on her wardrobe though, and that's one of the things she liked about him, his unassuming presence and just how down-to-earth he seemed. They didn't make a lot of men like that in Toronto, or at least, none that she had found. The way he looked at her, and paid attention to her before anyone else, Richard also really seemed to signal interest in her.

Back in her room, she dumped the contents of her backpack across the bed and realized she had packed a fresh pair of new yoga pants and a nice black top in case they had any dinners in town. "How about this, Ella?" she said, holding them up to Ella, to which she replied with a sideways cocked head and a loud half-grunt, half-sneeze. "Well, it's all I've got here, so it will have to do. He probably doesn't care what I wear anyway."

Anna quickly rinsed off in the shower, changed, ran a comb through her hair, and put on the only makeup she had: tinted lip gloss. She was ready in ten minutes. While she waited in her room, she decided to research some restaurants in town that might be good options for a quiet lunch. Before heading out, she gave Ella a quick brush with her comb, to which Ella squinted her caramel-colored eyes in pleasure.

Heading to the lobby five minutes early, she was surprised to see that Richard was already there waiting, browsing on his phone. "Hey Anna, I've just been looking up some places to grab a bite to eat, reading some reviews online. Do you have any preferences?" he asked in his beautifully thick accent. 'He is consciously early and a planner! Are there any bad qualities about him?' Anna thought with a wide smile.

"I'm easy! I mean, anywhere you feel like having lunch, I'm up for checking out," Anna piped up.

"Great, there is a top-rated Argentinian place called Milanesa on the main strip that we could check out. It's got a huge patio, and since it's so nice and sunny, if it's not too cold, we will get a clear view of the mountains," Richard suggested.

"That's perfect! Let's go check it out. Thank you for looking up places for us; I appreciate it. It's also less of a hassle eating with Ella on the patio. Most places with patios are dog-friendly," Anna said as Ella hopped cheerfully at the sound of her name as they made their way onto the street outside.

THE TOWN EVOKED feelings of an old western movie, with wide streets and rusted-out tractors, wooden hand-painted signs hanging from chains above restaurants and bars. It didn't take long for them to arrive at Milanesa on the main street, a giant log cabin building with a huge wooden deck facing the Patagonian mountain range.

The sun was burning high in the sky, but the air was still cool and crisp and smelled faintly of a wood-burning fire, a smell that evoked memories of Anna's childhood. Once they were seated, and they had ordered their lunch, Richard turned to look at her in a familiar way, green eyes locked on her, "Anna, you're an amazing woman, you know that right?" Anna didn't like the sound of his tone immediately and knew there was more to come.

"Well, I like you, but I live in London and I travel constantly and don't have any plans of settling down anytime soon. I just really like you, and I don't want to hurt you. I thought it would be kindest to let you know that, but let me reiterate that I like you. It just wouldn't be fair to either of us," he continued awkwardly.

Anna's face must have betrayed her attempts to stay cool and collected in the face of yet another letdown, this time before anything had even started. 'Why on earth do I even hope for anything? It all ends the same way. Every single time,' she thought as tears started to well in her eyes and she blinked them back fiercely so he wouldn't notice she was upset.

"Anna, are you ok? Did I just make a mess of everything?" Richard prompted.

'Did you make a mess of everything? What the heck do you think, Richard? Finally, a man who overthinks things to death more than I do,' Anna thought, totally dejected and feeling like she needed to get up and run away again.

"Yes, ok. That all makes sense. I just have to go wash my hands before the food gets here," feeling flushed and disappointed, she quickly excused herself to use the bathroom, and as soon as she was out of sight, the tears fell unabated. Ella ran after her and joined her in the bathroom as she cried.

"Why do I do this to myself, El?" Anna said out loud as she hugged Ella, her tears falling into her soft fur, "I give up. I don't even know why I try to put myself out there anymore or expect anything from anyone. It inevitably ends in heartache. You're the only one who has ever been there for me, and don't think I'm not eternally grateful." Anna tried to pull herself together in front of the bathroom mirror, splashed cold water on her face, and breathed deeply. She wanted to be sure she didn't look like she was crying; she didn't want to give Richard the impression he meant more to her than she meant to him. "Ok, hun, we got this," she said to Ella as she walked back to her seat.

"Anna, are you sure you're alright? Listen, I'm sorry if I said something stupid. I wasn't expecting to meet anyone while I was traveling," Richard looked concerned as she sat down and took a long gulp of her beer.

'Well, I'm not going to sit here and try to convince you to have feelings for me. After what happened in Guatemala, I'm not chasing down anything that isn't meant for me. I'd love to be chased for once,' Anna thought.

"Hey, I'm fine, really. Don't worry about anything; we're on vacation and having fun. Are you excited about the kayaking trip this afternoon?" Anna tried to change the subject. It was clear he was trying to push her to the side, and she wasn't about to put any more of her heart out again. She had been hurt already so much in the past year, and again on this trip that was supposed to be a recovery period. She decided that she would go back to focusing on her and Ella.

"For sure, yeah, I love kayaking. Ok, well. How is your dinner?" Richard asked awkwardly.

"Well, I haven't tried it yet, but the beer is crisp and refreshing, and delicious. I can't wait to go kayaking this afternoon. I kayaked back home all the time on Lake Ontario, during the summer, until my kayak was stolen, that is, and Ella used to come with me when she was a Miniature Schnauzer. It was her favorite thing to do," Anna started nervously filling the air with dialogue, trying to change the subject.

"Ella what?" Richard asked, puzzled.

"Nothing, never mind. I'm really tired and not thinking straight right now, ok? I'm sorry," Anna said in a tone indicating she was done with the conversation, as she moved to concentrate on her meal and gratitude for what she did have. 'Ella, you're the only one who ever really got me anyways, I'm sorry I was looking elsewhere for love for even a minute,' she thought, her gaze fixed on Ella.

After the meal, as the three of them were walking back to the hotel in awkward silence, Anna wanted nothing more than to get away from Richard. His mere presence reminded her that she had once again hung her hopes on something that wasn't real. She needed to grow up and stop fantasizing about romances that would never come to be. 'What was all that about manifestation anyway? I wanted someone to care about me and want me. And it didn't work. Big surprise,' Anna thought for a moment, feeling sorry for herself.

When they reached the hotel, Anna promptly excused herself, "I need to quickly go walk Ella before the kayak trip and get us both organized, so I'll see you in a little bit. Thank you for the lunch company." 'Yeah right, thanks for stomping on my heart before I even gave it to you. That must be a new record,' she thought.

"Can I come with you, Anna?" Richard called out. Anna pretended she couldn't hear him and kept on walking full speed ahead, needing nothing more than to forget what had happened and move forward.

Anna made her way back to the hotel just as the rest of the group was starting to congregate around the small bus for the short ride that would take them to the Río de la Vueltas, where they would kayak below the Fitz Roy Mountains. Anna and Ella would be sharing a single kayak, and she was glad that they would have this time alone to decompress after what happened at lunch with Richard.

THE TURQUOISE WATER looked like a sheet of glass in the still air. Colorful kayaks were laid out in a long line at the shore along the rocky beach, with the surreal backdrop of the snow-capped Andes. Anna selected an orange kayak that reminded her of the one that had been stolen from the National Sailing Club a year earlier. Just like old times, Ella hopped gingerly into the front of the kayak, panting in excitement, ready to go. "Is it ok if we push off?" Anna called out to Juan as the others were still getting their life jackets on and selecting kayaks.

"Sure, Anna, just don't go too far without us," Juan called back.

Anna pushed the kayak out, stepping into the water up over her ankles and noticing immediately how achingly cold the water was. By the time it was halfway up to her knees, the kayak was buoyant enough, and she got in, pushing the kayak the rest of the way out with her paddle. This kayak wasn't as wide or deep as her old one, but it felt comfortable and familiar. The sight of Ella once again at the bow brought an intense joy, filling the wells of Anna's heart.

"You remember this, right, baby girl? Your favorite thing in the world!" she called up to Ella, who turned around smiling at her. Anna's seat felt low and uncomfortable so she took her life jacket off and sat on top of it to get into a better paddle position. She'd been kayaking for many years and was an excellent swimmer. Ella had her adorable life jacket on, bright red with a large handle on the top, thanks to the kind gentleman at the rental center who happened to have a young golden retriever who was about the same size as Ella.

GLIDING ACROSS the turquoise lake, carving soft ripples in the water, it was incredibly calm. The air was silent and fresh, and the cold air brushed their faces as Anna gently paddled them out further onto the lake. With every paddle, she breathed deeper, feeling nature's energy passing through her body and feeling more at peace. This was what being on the water did for Anna, gave her a total sense of relaxation and peace of mind for which she was grateful after her painful lunch date.

Anna could feel the looming mountains calling her, and as she gazed at their snow caps, she felt their enormity dwarfing any problems she might have. She saw a group of black birds swim past the front of the kayak, and Ella in her usual instinctive way started jumping wildly and barking at them in excited greeting. Anna smiled happily, reminiscing about how her Miniature Schnauzer version of Ella did the same thing, only it wasn't the same. Ella was bigger now, and as she felt the narrow kayak flip, there was nothing she could do to stop it.

The kayak edge banged her hard on the head, and Anna was knocked unconscious in the water. She started to drift down into the reflected silhouette of the mountains, the blackness and the space between them enveloping her consciousness. Falling, sinking, deeper into the cold empty space, weightless. "What do you want, Anna? Do you even know anymore? The Existent Spirit loves you; God loves you and is a part of you more than you could ever comprehend. Do you want to be wanted, Anna? God wants you. Ella wants you. Do you want yourself, Anna? Do you see the immense beauty and power you possess?" The elder's voice came booming from the depths as the air in her lungs was replaced by frigid water and she was jolted back into consciousness.

Kicking and flailing with all of her energy, she tried to fight her way out of the darkness, but her shoes were heavy, her clothes were waterlogged, and she couldn't fight anymore. Her arms, aching and freezing, got slower as she stared up at the glimmering sun shining and sparkling down through the water and illuminating the dark creature swimming towards her. 'Are you god?' was Anna's last thought as she reached out and touched soft wet fur and felt strong teeth gripping her wrist, losing consciousness again.

"Come on, Anna, breathe, Anna! More chest compressions. Harder! Come on, Anna! Breathe!" a familiar voice came through, and she saw stars. Just black with specs of white shooting into the periphery of her vision, pain in her chest, intense pressure, and a wrenching cough. Anna felt extreme pain in her chest as she coughed up water; her lungs were on fire. "Yes, yes, turn her on her side and drain her lungs," the voice again, as she felt pounding on her back.

Anna came back as Ella jumped on her legs, nipping at them, pawing her vigorously, then running up to lick her eyes and face. She coughed, in terrible pain, and couldn't move. Juan's face came into view as he placed a blanket around her and propped her head up with a life jacket, "An ambulance is on its way, Anna; you're going to be alright now, thank God, and your precious dog, you're going to be ok. She tore that life jacket to shreds trying to get it off her though. Good girl Ella!"

FAITH   FEAR   FORTUNE

ANNA WAS RUSHED to the small nearby hospital, just twenty minutes away. "Yes, sir, she is going to be fine. She just has a mild concussion and we have had to drain some fluid from her lungs, but overall she is in great shape and incredibly lucky. Nice work on the chest compressions. Your quick thinking and actions surely saved her life," an unfamiliar voice, and a sterile smell. All Anna could see were the tops of robin's egg blue walls, a speckled ceiling, and fluorescent lighting that was far too bright.

"It was nothing, besides that, her dog is the one who got her back to the shore so fast so I could even perform effective cardiopulmonary resuscitation," a familiar voice, and the smell of warm campfire mixed with a spicy aftershave beside her, accompanied by a rough warm hand on her arm.

Feeling a rough wet tongue cleaning the bottoms of her exposed feet, Anna opened her eyes wider and lifted her head, "Ella! Baby girl! What happened to me? Where am I, what's going on?"

That familiar voice again, which she now recognized as Morten's voice beside her, "Anna, dear girl, you almost drowned. If it wasn't for your brave, brilliant companion here dragging you back to shore, you might have died on us. Ella saved you, Anna. She saved your life, thank God. How do you feel?"

Richard, who she now noticed was sitting tucked away in the corner, added, "Well, Ella and Dr. Morton here who was able to perform CPR on you and get that water out of your lungs. Fortunately, we had the talented Dr. Morten with us. Thank God. I am so glad you're still with us Anna; you gave us all quite a scare, me in particular. Yes, enough from me, how are you feeling now?"

Anna tried to talk, and her throat hurt and burned as she spoke, "Honestly, I feel like I've been run over by a freight train, but after hearing what you told me, I'm just grateful to still be among the land of the living," she managed as she rubbed the top of Ella's soft head propped up on the edge of her bed.

"You're going to have to stay in the hospital for the next two days; it's policy. They will keep you under observation to ensure there are no secondary lung infections or other complications. You should be feeling better in a matter of hours though. You're very fortunate, Anna," Morten said again.

"That's really good news. I don't like being in hospitals, so the sooner I can get out of here, the better," Anna felt relieved, "Will I be able to rejoin the group after the two days I need to spend here?"

Morten thought about it for a minute, "I don't see why not. If you're feeling up for it and your breathing isn't affected, it should be fine. It will depend on how you feel and how quickly you can make a recovery, but we would obviously love to have you join us for the hike to Los Glaciers on Friday. Just focus on resting for now, and play it by ear."

"Ok, that makes sense. I've made it all the way to the edge of the world, and it's taken me so many years to get here. It would be a shame if I couldn't fulfill my dreams and hike the mountains of Patagonia that are so close, right there outside my hospital window," Anna said with determination, "I will focus on those mountains, and I will recover quickly."

"You are indeed a remarkable person, Anna. I have no doubt that you'll be back with us on Friday, and at the front of the pack," Richard encouraged her.

"Well, actually, I will be at the back of the pack, but I quite like it there," Anna laughed at herself alongside chuckles from Morten, Richard, and the two nurses that were crowded into her tiny room.

"Ella and I will be on either side of you," Richard said, smiling warmly, "Now go back to sleep, rest, eat and get yourself better."

As everyone left the room except for Ella, Anna marveled at the miracle dog who had somehow ridden the Existent Spirit energy back to find her. Love, it seemed, was truly stronger and more powerful than anything else, and it had saved her life. "All I ever wanted was to be loved, and I am. I am so grateful to you, Ella. I love you so much," Anna said to Ella through tears of joy, "And I do want myself; I want to live. I'm not afraid of life anymore. We will do it together."

Ella jumped up and curled herself at the end of Anna's bed to keep silent watch as she slept. There was nothing that would be able to stop their love for each other. Not even death.

AMY C. SHEA

Los Glaciers National Park, Patagonia

19

The Love

Anna spent the next forty-eight hours, as the doctors recommended, resting with Ella by her side. She wanted to be in top shape for the hike that they would take together through the mountains. All the time in the hospital gave Anna plenty of time to reflect on her life. She had spent years looking for love and companionship in someone else, a love that had always been right there in front of her and had never left her. There had never been a love like the one she and Ella had for each other, and she was eternally grateful that somehow she had found her way back to her.

Through all of the pain, trauma, and loss, Anna had also somehow found a way to face her deepest fears and had pulled courage from places she didn't even know existed. She was proud of herself, for the first time in years. She had found a way back from the depths of despair and had chosen to keep living when she had only seen death all around her. Had all of the terrible, traumatic things not happened, Anna would not have had the chance to experience her resilience.

"Ella, how is it that the worst time in my life not only brought you back to me but showed me who I am again," Anna said out loud to the beautiful creature at her side. She closed her eyes and prayed in gratitude, thankful for everything, good and bad, that had brought her to this place. She was stronger, wiser, and finally at peace with the woman she had become.

Anna had somehow found a way back to herself after being lost for so many years. She didn't understand the path she was on, but she had chosen to have faith and keep going, and that was the most important thing. It had all led her to this place, where she would be facing one of the toughest physical challenges of her life in the Patagonian mountains. The old Anna would have been terrified, but she wasn't afraid. She had Ella, and she had her confidence back. Together they would be able to conquer any challenge.

THURSDAY MORNING came quicker than she had imagined, and Anna was discharged from the hospital. She was surprised to see Richard had come to meet her and was waiting in the lobby of the hospital, "Hi Anna, Juan was going to come and meet you, but I asked him if I could come instead. I feel bad about the other day, and well, I was hoping I could apologize."

"It's ok Richard. I understand, and there are no hard feelings, ok. Thanks for coming to meet me; it's really nice of you," Anna was happy to see a friendly face even if Richard had upset her the other day.

"I was getting ahead of myself the other day, and it was foolish. I don't want to get hurt, Anna, and well, the feelings I'm developing for you are real and it scared me. It's been a long time, and then when you almost died the other day I realized how stupid I had been," he continued.

"I didn't almost die; stop being so dramatic," Anna laughed.

"You were turning the color of a cranberry, Anna. I love cranberries but that was just disturbing. Anyways, can we just go back to getting to know each other and forget the colossal blunder, and the mess I made of things?" Richard asked.

"I think that's totally doable. I'm not going to lie, I was hurt the other day because honestly, Richard, I was, and I am starting to like you too. I know it's not the most convenient situation, but yeah, there it is. If nothing else, at least we have the start of a great friendship," Anna smiled at him.

"Well, I truly am sorry that I hurt you. That was the last thing I wanted. To make it up to you, can I at least buy you lunch for ruining our last lunch?" Richard offered.

"Yes, of course, that would be wonderful. It's been two days and I'm already sick of hospital food, and we are all going to need our energy for the hike tomorrow," Anna agreed to the lunch date, and they went to a Seafood Barbecue three blocks from the hospital before walking back to the hotel.

"Thanks, it was nice of you to come pick me up and take me to lunch. No matter what happens, I'm grateful that we met. I'm going to go get packed and get psyched up for the big hike tomorrow,"

Anna hugged Richard warmly and went inside with Ella to prepare for the hike, "See you tomorrow morning, Richard!" she called back.

"See you tomorrow, Anna. Rest up," he called back as he continued down the street.

IN HER HOTEL room, as she got ready and started taking inventory of the items she would need for the hike, Anna thought harder about starting any kind of potential romantic relationship with Richard, and whether that was even what she needed. Even if she couldn't speak, Ella was a great listener and seemed to guide her decision-making with just a blink of her eye or a tilt of her face.

"Ella, when I started this journey, I was trying to heal from a terrible relationship and running from the pain of losing you. I have you back now, but I don't think I'm done healing from all of the pain that I went through. I like Richard, but it doesn't seem like he knows what he wants, and I can't afford to get hurt again right now," Anna reasoned with Ella, who acquiesced with a gentle nod of her head and intent eyes.

"I think maybe I'm ready to trust the Existent Spirit and let God show me the way. When I started this journey, I could manifest your energy just by challenging myself in nature. Scaling Fuego, climbing the ruins of Tikal, running up the Hill of the Cross, and The Lost City Trek, these were all incredibly difficult things in the heart of the wilderness that pushed me to the edge physically and mentally and connected me to you through the Existent Spirit," Anna continued to Ella, who just grunted.

"So even when I don't know where I'm headed, I just need to keep challenging myself. I think that's the point of all of this, to just get up and keep going, to trust and to have faith that this journey is the point of all of it. When I was sinking into the depths the other day, the elder Ezquierda asked me what I wanted, and it was to live. I want to live my life and not be afraid to do it anymore. I've never been so excited to live my life," Anna said, hugging Ella close to her in gratitude, "You, my El, are a precious miracle."

Ella sighed in Anna's arms, and then after Anna had let her down, she curled into a cozy ball in the corner against the pillow at the top of the bed. "I love you, Ella. I always believed that love and our spirits couldn't be destroyed. Thank you for proving this to me," Anna said good night as she lay down on the pillow beside Ella and drifted off to sleep.

20

The Shared Belief

The sun shone its warmth in geometric patterns over the walls of the small hotel room as it rose at five-thirty. Anna sighed loudly and rolled Ella over to spoon her and give her a huge cuddle before rolling much more slowly in the other direction out of bed.

"As much as I would love to get a few more hours of sleep, adventure awaits, life awaits, and surprises await us, Ella! Get out of bed, lazy head. Let's get some breakfast," Anna sang to Ella to get her up as she hopped in the shower and turned on the cold water to jolt herself into action. Ella stretched her front paws out on the bed and then slumped back down with her head resting on them, aiming to get as much downtime as possible before Anna dragged her out of bed.

Incentivized by the excitement of exploration, Anna put together her last pieces of gear and stood ready at the door like a real mountaineer. There was a lot to bring because they were told to expect any weather conditions they could imagine. "I feel like one of those mules in The Lost City, El," she exclaimed. Ella responded by jumping onto the floor and grabbing the fabric of her pants gently with her teeth and pulling.

"Yeah, ok, I know it's early. We should go and get some coffee and fuel up though. I hate rushing through breakfast in the morning. Let's go," she replied, pushing open the door so Ella could lead the way up to the breakfast area. Ella scampered clumsily up the stairs and bounded into the arms of Richard, who was already sitting at the table with Juan.

"Anna, good morning to you and Ella! Are you two ready for the thirty-kilometer hike today? Make sure you eat a lot and get your energy stored up," Juan sang chipperly between gulps of coffee, "We're going to show those mountains who's boss today! I'm so glad to see you've made such a fast recovery, but remember to go at your own pace. It's going to be a hard, steep climb at times."

"Good morning, Juan, happy to be back in action. Don't worry about me; there is no chance I can go faster than my own pace, trust me. Ella knows how glacial a pace I keep. Come to think of it, the glaciers might move quicker than I do," Anna joked back.

'Wow, Ella sure does like Richard,' Anna thought to herself as Richard smiled warmly at her and ruffled up Ella's fur with his hands, 'But that doesn't mean that I need to get drawn into some intense drama again. He's a nice guy, but we will leave it at just friends. You can only get so hurt by your friends; romantic relationships are a different story.'

Anna opted for a large plate of scrambled eggs with three pieces of toast with butter, sweet potato, a cheese empanada, fruit, orange juice and coffee. As she loaded up on fuel for the hike, the rest of the group came filing into the breakfast area laden with gear, hiking poles, down layers, and packs. Ella was finishing her bowl of food, and Anna gave her some extra egg and a few large pieces of sweet potato before getting up for a quick walk.

"Is it ok if I leave my things here while I take El out for a quick walk before getting on the bus?" Anna asked the group.

"Yes, Anna, we will be here!" Morten said smiling through teeth full of toast, "So glad you're feeling better, and happy to have you on the hike. Just watch your lungs, if your chest starts hurting at all, stop and rest."

"Thank you, Dr. Morten. I appreciate everything you did for me the other day, you know, saving my life and all." Anna owed her life to Ella and Morten.

"It's nothing, Anna, and please, it's just Morten to my friends. Only my wife calls me 'Dr. Morten," he said coyly, with a chuckle.

THE DAY couldn't be more picturesque. The air was fresh, but not too cold, and there wasn't a cloud sullying the rich blue skies above them. That familiar, faint campfire smell hung in the air as Anna walked Ella to the parkette across the street from the hotel as squirrels scurried out of their way.

Walking back, Anna saw the group in front of the hotel and the bus had pulled in early. Starting to jog back, Richard yelled, "No rush, I grabbed your things already, and they're on the other side of the bus. Also got some extra treats from the kitchen for Ella."

"You didn't have to do that; thanks, Richard, that's very thoughtful of you," Anna yelled back. 'I hope he doesn't still feel guilty about the other day. How could I ever fault someone for being scared and confused, I've been living in that state for the past year. At least he was honest with me,' Anna thought to herself.

"He sure knows how to treat you, Miss El," Anna said in a low voice to Ella as she trotted back in front of her, smiling teeth and sparkling crescent eyes occasionally shooting glances back her way. Ella hopped on the bus before anyone else and curled in the front seat.

"Sorry, Juan! She knows where she likes to sit," Anna said embarrassed as she followed her and sat down.

"No worries, Anna. We only have about an hour this morning before we get to the trailhead. We're going to split into groups because I don't want anyone hiking this trail solo. It's not anything like most North Americans or Europeans have experienced before; it is very challenging and can be quite treacherous as well, mostly due to the weather," Juan explained as the rest of the group was boarding the bus, "There was a man in his twenties with a friend last year and he fell and succumbed to his injuries."

"He didn't survive? That's terrible news. So young to have passed away hiking." Anna asked to be sure she understood what she was hearing.

"Yes, he died. He fell backward and cracked his head on a rock. It rained and the rocks froze and were slick. I take the safety of our groups very seriously because accidents can, and do happen often on this particular trail route," Juan continued, "It's not usually about skill or athletic ability, sometimes just an accident, usually related to the rapidly shifting weather. This is why we need to stick together."

Anna was glad to have her trusted hiking sticks and sturdy boots, and Juan's words were etched like a stern warning and burned into her mind for the rest of the bus ride. She wasn't nervous but she was determined to go slowly, carefully, and heed his warning. Anna also took comfort in knowing that Ella was young-bodied and sure-footed with lots of experience hiking in Colombia. She didn't have to worry about her at least.

AS THEY MADE their way across the plains towards the mountains, clouds started to roll in from the east and obscure the sun. Looking past the mountains, Anna could see streaks of rain painting diagonal lines through the encroaching white clouds. 'Literally sunny one minute and rainy the next,' she thought to herself, happy that she'd packed all of the gear that was on the list the tour company had provided. By the time the bus reached the trailhead, the gray rain had completely rolled in and the tall white peaks were obscured by rolling misty clouds. Stepping off the bus, the damp air felt cold against her cheeks and she pulled down on her wool cap. "That's an enviable coat you've got there El," she said as Ella jumped off the bus and shook herself out in the light rain proudly.

Ella had always hated being wet. When she was a Miniature Schnauzer, Anna had often taken her to the beach on her kayak. If her face got wet she would try and wipe it off on the sand, which only succeeded in firmly embedding wet sand up her nostrils and in her fur. Grunting in frustration she would zip around the beach, pouncing and barking at the waves. It didn't seem like there would be much either of them could do about the rainy weather though. "Do you want your rain shell?" Anna asked, taking it out of her backpack, to which Ella responded by taking off up the trail. "I think I'll leave this on the bus then," she said, balling it up and throwing it under the seat at the front.

Juan assured her, "Good call, we're hiking for over ten hours today so you don't want to be carrying anything that you don't absolutely need," then louder so the rest of the group could hear, "Did everyone hear that? We can leave anything on the bus we don't need right now. Steep rocky inclines mean, the less weight the better."

As they started to hike, Anna quickly caught up with Ella, who was waiting just ahead and was sniffing around at the back of the visitor center, likely reading pee-mail from other tourist dogs. "Let's go, my hairy, seasoned hiker. Time to hit the trails," she called to her, taking intentional strides forward at the back of the pack.

"Do you mind if I hike with you?" Richard said as he caught up to her and Ella from his bathroom break at the visitor center.

"Yes, of course, but I warn you that I'm going to be taking it extra slowly and I don't want to hold you back. I'm a bit worried about pushing my lungs after the other day and I do have a trail mate already, after all. Although she is pretty flatulent today so I might not hike behind her. Come to think of it, you probably caused it with all of the eggs you were feeding her," Anna motioned to Ella laughing.

"Duly noted about Ella, I'll give her some space. I don't need to be at the front of the pack; I'd rather just take it easy and enjoy the journey. The scenery is why I'm here, Anna, and the slower we go, the more time there is to enjoy the beautiful Patagonian vistas. I also hate cheesy handheld selfies and wouldn't mind having my photo taken by you a few times if I'm being fully honest."

'He hates selfies. Ok, I'm starting to like him even more,' Anna thought, laughing. "Ok, let's do this then. I still can't believe that I'm in Patagonia. A few months ago, I was feeling all down and depressed in my condo in Toronto, and now I'm climbing a real mountain!"

"I hope you can tell me why you were down and what's happened in your life. I'm here to listen if you want to talk to someone. I can be a really good friend, Anna, I've got a sister and we've always confided in each other," Richard offered.

"Sorry, I didn't mean to open up a can of worms on you. It's just been a really hard year, and there has been a lot of loss. There has also been a ton of growth that's come out of it, and I've been equally blessed. I feel stronger than I ever have before, to be honest," Anna continued.

"I am sorry to hear about your difficult year. Mine has not been the greatest either; this is part of the reason I'm taking this trip, to have time to process the death of my brother. He was killed in an auto accident last year, and it was really hard on my family, and that's partly why my wife and I got a divorce six months ago," Richard told her.

"Richard, that's terrible. I'm sorry to hear your year has been so rough. I'm glad that you decided to take this trip for yourself. It sounds like you really need it, and it will be good for your spirit. I'm really sorry to hear about your brother's passing, I can't imagine what you've been through," Anna said as she started to feel a ton of empathy for him. It turns out she wasn't the only one having a difficult year.

Richard continued, "The divorce was a long time in the making. We were terrible for each other, but after being together for so long sometimes it's just easier to keep going along with things. It was easier at least until my brother died, and I got a wake-up call that life is too short to be miserable. My brother just wanted me to be happy, and I hope that even though he isn't here anymore to see me making a shift in my life, that he is proud of me. I decided to take this trip to go on the adventures he always wanted to take but never got the chance to experience. I also quit my job in finance and I'm following my passion, writing."

"I'm sure he is very proud of the life you're living. It's really easy to just give up and keep going along as you say. What takes real heart is to keep living when all you want to do is curl up in a ball and give up. It doesn't matter how you move forward; what matters is that you are moving. Not only are you moving forward, but you're courageously making changes in your life to set you up for future happiness. You're inspiring, Richard," Anna assured him.

"That's some pretty sage encouragement, you know that? If you ever want to talk about your rough year, I'm ready to listen. Any time, Anna. I feel like I can open up to you, and that's something that's always been difficult for me," Richard said, thanking her as they continued to climb the path that was getting steeper.

Anna was starting to understand Richard's apprehension about developing feelings for her. She felt bad about rushing to judgment earlier. Richard was hurting just as badly as she was, maybe more. It was pretty clear he wasn't toying with her at all, but rather, he was afraid just like her. They were more alike than she knew.

"Thanks, Richard, I will probably take you up on that one. I feel like I can talk to you too," Anna could barely manage, already out of breath in her attempts to match Richards' long strides up the path. Ella, full of beans, was yards ahead of them both, bouncing over the boulders as though they were made of rubber.

FAITH   FEAR   FORTUNE

BY MID-MORNING, the rain had subsided, and the sun had come out. The strenuous hiking, combined with the sun made Anna feel warm, hot even, and she'd taken her down coat off and packed it neatly away. The vistas were incredible. She could Google all she wanted but the immensity and visceral intensity of the landscapes could never be captured or described in even a thousand Google images. It was the feeling of being dwarfed amidst the enormity of a black hole.

The feeling of being minuscule in the face of nature's awesome scale gave Anna a hopeful feeling, knowing that her reverence for the natural world, the Existent Spirit, and its power would always be far greater than her mortal body and mind. There was so much she didn't understand, and this gave her comfort. The vastness moved her to tears.

"Anna, are you ok?" Richard asked.

"Yes, more than ever before. Can I ask you a question, Richard," Anna said nervously.

"Of course!" he said quickly.

"Do you believe that there's more to this life than just the physical world that we can see? Just the mountains, our bodies, our flesh, and our brains?" Anna asked, staring out.

"You know, Anna, if you asked me that question five years ago, I would have said that I hoped there was more. Now, however, I know one hundred percent that life goes on. I can't prove it in a scientific way but I'm sure that life goes on. Life doesn't end when our hearts stop beating, and I've had the proof shown to me," Richard said confidently.

Anna just stared at him, shocked and delighted at his certainty, for a few seconds, then managed, "How is it that you're so sure of this, Richard?"

"I'm going to preface this by saying that I never used to be a religious or spiritual man. I told you my brother died, right? His name was Harry. Well, when Harry and I were kids, we played this weird game where we would take our favorite toys, and when our parents were sleeping, we would stage them around the house like they had gotten up and walked around in the middle of the night by themselves. My brother would take his Star Wars Chewbacca figurine and get him into all sorts of mishaps, 'Hijack Harry' is what my parents would call him," Richard reminisced.

"Ok, and what does a Chewbacca doll have to do with my question?" Anna laughed.

"The best part is coming Anna," Richard responded intently, "That was thirty years ago now, that we played that game. We were ten. Well, after he passed away last year, someone sent me a package in the mail. It didn't say who it was from. I opened it up, and it was the darn Chewy doll, Anna. I'm serious. You can't just make this stuff up, I was dumbfounded," Richard said incredulously.

"I believe you. I really do," Anna was starting to get chills up her spine.

"Well, I didn't know what to do with what happened, so I just chalked it up to someone thinking about me and getting me a present. Someone must have known Harry and I used to play with that doll, and they wanted to get me something comforting from my youth without wanting to take the credit. I know that sounds ridiculous, but you tell yourself stories so you don't feel like you're losing it, Anna," Richard continued.

"You might not believe me, but I know exactly what you mean. So what else happened?" Anna asked, sensing there was a lot more to come.

"If that wasn't strange and coincidental enough, I put the doll on my desk, assuming that someone I know would eventually fess up to sending it, but that never happened. About three months after Harry died, Chewy started disappearing off my desk for days on end. I thought I was going nuts. My ex-wife would find it in the fridge or the garage and accuse me of doing it in the middle of the night. She made me go to therapy, and then a sleep clinic, and she wouldn't buy my theories behind what was happening. It was Harry. I'm a thousand percent sure he did it to comfort me. I didn't tell anyone about our game; only my parents knew, and they're both deceased," Richard finished.

"Well, you're not nuts. I'm so sorry, Richard, about your parents, about Harry, about everything. It sounds like all you needed was some support and someone to listen to you and believe you. Well, I believe you, Richard. I one thousand percent believe that it was Harry bringing you comfort from beyond this world. He loves you so much, and love like that never goes away. I'm also sure of it. I have my reasons, and proof, but trust me, I believe you," Anna assured him that he wasn't the only faithful one on the mountain.

"How do you know, Anna? Why do you believe that?" Richard prompted.

"Well, I'd say that you're going to think I'm nuts, but after your story, I think you might believe what I'm about to tell you too. So here goes. I have always believed in God, I just didn't know what God was, exactly. Well, Ella and I have known each other for about seventeen years now," Anna said carefully.

"But she's a puppy. How?" Richard asked incredulously.

"Well, I didn't have proof of a life beyond what we can see, and then when my ex committed suicide in front of me last year I got my first real proof. Calvin dragged me to the balcony, and I felt a hand stop me, on my chest, so hard it knocked the wind right out of me, and I fell. That hand saved my life from death that day. I don't know if it was God or what it was, but in that instance, I believed. At that moment, I wasn't alone," Anna confessed.

"I'm so sorry, Anna. That's terrible. But I'm so glad you're alright," Richard said sympathetically.

"That was just the beginning. After I witnessed the suicide, I was diagnosed with post-traumatic stress, and I couldn't work anymore, and then a few months later, Ella died. I thought my life was over, and I didn't want to go on. I was so depressed. Ella was a Miniature Schnauzer in her last lifetime, and after that body was done, and she was free, she started to send me signs to comfort me. The signs pointed me to South America where we were reunited again in The Lost City. I know how this all sounds, but it's true," Anna finished the abbreviated version of her story.

"I did wonder how a three-month-old puppy was so well-behaved. It does sound far-fetched, Anna and the old me would have laughed at you and your story, but I believe you. I do believe you. Not everything in this world can be explained. I know that now. It's kinda like how nothing is really solid, it's just an illusion, and we're made of subatomic particles. This is how I see faith. It's something you can feel, even when you can't see it," Richard set Anna's mind at ease.

"Thank you for not thinking I'm a nut job, Richard. All of the obstacles in the past year have challenged my limits around what I believe. If nothing else, I've been fortunate to have my faith renewed. I don't know what God exactly is, but I'm starting to understand that God is nature and the Existent Spirit that flows through the natural world. I believe that we are all connected and are one," Anna said.

"Existent what?" Richard asked.

"The Existent Spirit was introduced to me through my struggles, but I didn't understand it until I spoke to an elder named Izquierda of the Arhuaco tribe in Columbia. He told me that by tapping into the Existent Spirit that binds us all together and unites us, we can manifest what we need, and reunite with our loved ones. It is what drew Ella back to me, and it's what keeps your brother close to you, I'm sure. At the core, it means that you and I are part of one another, and we are therefore already whole," Anna explained.

"It sounds very Buddhist. Reincarnation of our spirits and one whole that we all belong to. I've always been drawn to Buddhism and the notion that we are all connected spiritually. That all feels like it makes sense to me," Richard said.

"Yes, it felt true to me too, and then when I experienced it in The Lost City, I knew it was real. I also felt deep in my soul that all religions are interpretations of what I felt. No one is wrong, and everyone is right because there is one God and one spirit, The Existent Spirit. All of the wonderful things that Christ did, were done by God to prove to us that we are loved. Hinduism, Islam, Catholicism, Judaism, at the heart it's all about love. The interpretations are what man does to twist religion for his purpose, to fulfill greed. You are a part of me, your brother and Ella, and everyone, regardless of their religion. Do you understand?" Anna continued, unsure how this knowledge from the elder Ezquierda was coming to her.

"Yes, I do. Thank you for sharing all of this with me, Anna. I think there's a reason we met and are having this conversation on the mountain. If what you're saying is true, then perhaps my brother brought me to you to reconfirm that his message is real and to bring me further comfort. He also really wanted me to be happy. Meeting you, Anna, it feels like it was meant to be," Richard said, smiling warmly.

"Pretty deep conversation for a Monday, right." Anna winked at Richard.

"The best kind of conversation Anna," he smiled warmly back at her, feeling completely at ease.

Staring into the enormity of the Patagonian mountains, Anna and Richard were looking out in the same direction, and in a matter of minutes, they had gone from virtual strangers to a place of deep alignment and spiritual understanding. Anna had never felt more at peace standing next to someone or more understood.

21

The Guardian

The ranges of Torres del Paine stretched like giant hands into the cobalt-blue skies as they continued to ascend. Anna had to concentrate on each step over the loose mountainous terrain. She thought of how earlier in the week, she had sunk beneath the mountains into the glacier waters, and now she was moving over them. It reminded her that the most important thing in life was to keep going even when things seemed to be at their most bleak.

A few months earlier she had been in the hospital, devastated over the loss of Calvin. Even more recently she had thought she lost her best friend Ella. What gave her comfort was knowing that no matter how bad things could get, as long as she could take her next breath, there was hope. Nothing was ever static, even the monstrous glaciers were moving. Here she was, scaling the mountains of Patagonia with her dearest Ella back, and hope and faith gracing the deepest hollows of her heart. On top of all of that, she had made a genuine connection and a promising new friendship.

The path up through the mountain pass was getting treacherously steep, the air was getting thin, and Anna was starting to slow down.

"Richard, please don't wait for me. I'm going to be moving at a snail's pace for a while. I seriously need to slow down a bit; my legs are screaming at me. I've got Ella here," she said breathlessly, "Don't worry about me."

"Anna, Juan told us to stay together because it's treacherously steep. I'm a little annoyed that the group isn't waiting for us after such a stern warning on the bus. I'm going to go up and call them to slow down. I'll be right at the next bend waiting for you, ok," Richard said.

"Yes, that sounds fine. Like I said, I've got this. You take a break at the top, and we will meet you there," Anna said, stopping to catch her breath every few words, with Ella statuesquely perched beside her on a large boulder.

The frigid wind was whipping her backside as she continued to climb the steep rock face, looking up the terrain to see Richard disappear around the corner. She could hear him yelling toward the group. Small fir trees had somehow tenaciously grown their way into the cracks on the sides of the rock, and they reminded Anna of the small bonsai tree she had tended to as a child. She continued to climb steadily but carefully, and the rocks started to get more sheer, the incline more pronounced.

"Wait a second, El, I'm going to ditch the poles for a while so I can use my hands to steady myself. It's too steep for these right now," Anna said as she collapsed the poles and attached them to her pack with a carabiner just as she felt the freezing rain start to pelt her face.

"Ugh, why now," she exclaimed as she started to climb up and over the rocks which were growing slippery and wet. "I wish I had sandpaper grip paws like you, El," she said to Ella who bounded across the rocks with the sure-footedness of an alert, caffeinated mountain goat. Ella turned around and barked shrilly at her once and then kept moving upwards.

THE RAIN got heavier as the minutes passed. 'Climb up higher, just keep moving. It's going to be beautiful at the top, and just think of all the fresh air and exercise you're getting. Mind over matter,' Anna fought hard to keep a steady stream of positive thoughts flowing amid the physical pain. She looked up at what seemed like a torrent of water rushing down the rock toward her and gasped.

"What's happening? Richard! Ella!" Anna screamed terrified, gripping the rocks and steadying her feet so she wouldn't get swept down the path as a stream of water rushed down the mountain. She clung motionless and white-knuckled to the mountain face as the rain soaked and chilled her to the bone.

It seemed like the rush of water was over, so Anna kept climbing. All she could see was a shining silver path of water up the mountain, and Ella's long tail wagging back and forth at the top. She used all of her energy to climb towards it, her cold wobbly legs making it more challenging. Her hands were freezing and cramping up.

"Anna, you do know why you're here right?" It was a voice again, booming and filling the spaces around her. It was not the elder, and not Ella, because she was already here with her. It was a woman's voice, familiar, soothing, and warm. It comforted her like warm oatmeal with maple syrup on a cold Sunday morning.

"Who's there?" Anna asked as she stopped climbing and looked around.

"You know who this is. I've been with you this entire journey, before that even, and I've been trying to help you get in touch with me. It certainly has taken you a while though. You're just as stubborn as you were when you were five years old. Dear, I wasn't about to let that misguided man take you down with him. It wasn't your time to go," the voice came again, kind and soft.

"Are you going to tell me who this is? Do you mean Calvin?" Anna asked.

"You didn't imagine that metaphysical hand on your chest that day. When I passed from the physical world, all of the mysteries of life were revealed to me in a single moment as I returned to the whole. Just as you've been using your gift and manifesting what you need here on earth, I can manifest what I need through the Existent Spirit. I needed to save you that day, Anna. You have places to go," the voice kept speaking gently as though they were sitting together in a living room chatting over tea.

'So familiar and warm, who is this? I know this person deeply,"

Anna thought. "I didn't imagine that hand that day. I knew it happened! That was the one thing that seemed so real that day, and I couldn't explain it. There was so much anger; he wanted to kill me that day, but it didn't happen. You intervened. Who are you? Thank You, whoever you are, from the deepest wells of my heart," Anna pleaded for answers, full of gratitude.

"Remember when I told you that you were going to grow big and strong and have all of the wild adventures that I wish I could have taken? You needed to see the world and do the things that I dreamed of. You need to know that I'm with you every step of your life and going on these journeys with you. Do you think you're lonely? You could never possibly be alone, Anna. These are my hands; my hands will do a thousand loving tasks for you, and when I'm grown up and very tall—" the voice continued.

"—you will remember that my hands were once this small! Grandma Catharine! How? Is it really you?" Anna burst out in tears of pure elation as she finished the long-forgotten words of wisdom.

"Yes, it's me. Of course, it is. You can't possibly think that a dog's spirit goes on but a human's doesn't? We're all the same, dear. I know how much you loved Ella, and I made sure she found her way back to you. She's your animal spirit guide. These things happen all the time; it's just that most people don't have the gift of tuning into the Existent Spirit, so they are closed off to the natural world. Our loved ones are, in fact, all around us. There is no such thing as death; it's all just a continuum," Grandma Catharine went on.

"Then why don't people understand? This very fact and knowledge of life after death would end wars, division, and so much pain. If people only knew, we could save so many people from the agony of the human condition," Anna said through tears.

"Humans are the limited expression of that which is in fact limitless. Everything natural, every animal, plant, and organism has important lessons for us humans if we would only listen. If we listen, we will become a part of the Existent Spirit in this human form. Once you have awakened to it, there is no going back. You will seek to maximize human and animal happiness on this precious earth. It is the human mind that creates the illusion of separation, and we need to deprogram years of desensitization," Catharine continued.

"Is this why the elder Ezquierda told me to get closer to nature then? I've been so disconnected, even before Calvin's suicide. I didn't know who I was in this world anymore, Grandma. I could never understand my place in the world. I've only just started to reconnect to the spirit and the natural world and it's healing me," Anna continued.

"Yes, through nature, we can see that we are neither limited nor separate. War, pain, and heartache are a result of us seeing ourselves as separate and imagining that we have a lack of what we need or an imaginary expiration date associated with our spirits. Anna, you have the courage to tell others what you know in your heart to be true. Like the comfort you were able to provide Richard," Catharine continued.

"And it was true that his brother was coming to provide him with some comfort then, wasn't it?" Anna asked.

"Of course it was Anna. There are so many ways to communicate and send messages to our loved ones, and as I said, not everyone has the gift of being able to use their senses to tap into the Existent Spirit directly. Richard's brother used a symbol of joy and happiness from childhood that he would understand. We can manifest movement when we need to, as you learned," she went on.

"There is still so much I don't understand, yet my faith is rock solid. Thank You, Grandma. I always felt you with me. I love you, and I miss you," Anna felt intense joy and gratitude for what was happening.

"Your mind has opened, the change has happened. You are aligned with the Existent Spirit, and you will never need to miss me again. We can speak just like this whenever you like now. Ella needs you though, so keep climbing! I've always wanted to climb a mountain; now let's do it together!" Catharine pushed her on, and Anna could feel her grandmother enter the broken spaces of her heart and wash away all doubt as she continued to climb to the summit.

The rain subsided, and Anna felt a warmth pass through her body. The cold-induced cramps in her hands started to fade, and she pushed onwards with the comfort of knowing that life goes on.

FAITH   FEAR   FORTUNE

What great fortune had smiled down upon her and given Anna the surety that she would be alone again? 'Life is beautiful,' she thought to herself, grateful.

22

The Turn

A loud call came from up ahead, "Anna! Hey, I thought I heard you screaming. Are you okay?" Anna saw Richard's bright blue beanie as she rounded the corner and reached the peak.

"Yes, I got scared when the water started rushing over the rocks, but I'm ok now," Anna said.

"What water? I didn't see anything from where I was. Are you sure you're alright, Anna?" Richard was growing more concerned.

"I just had a very spiritual experience; you're not going to believe it. You probably will. I'm starting to see you're the kind of person who has an open mind," Anna said, relieved as she caught her breath. The thin air was starting to catch up with her, and it felt as though she was breathing through a constricted hose.

"Whatever it is, Anna, if you truly believe it happened, then I believe you experienced it. I'm starting to feel like all of this is predestined," Richard said, extending his hand to help her up the last steps, "Isn't it just incredible? Breathtaking."

The summit gave them a panoramic vista of tall rugged black and white mountains, and Anna had never experienced a moment as fully as she did just then. There was no past, no future, just the mountains. Everything was one, and also within her. Somehow she knew that it was all going to be ok, not just for her, Ella, and Richard, but for everyone she knew.

"I want this feeling of oneness and presence to be experienced by everyone. I've never felt so at peace, Richard. I finally understand happiness, it's being in nature with the Existent Spirit coursing through me," Anna turned to Richard, looking deep into his eyes which were also welling up with tears of joy as he smiled at her. Anna sensed a profound connection to him at that moment, one she had never before experienced in her life.

Pulling Anna to him, Richard embraced and kissed her softly on the lips. "I love you, Anna," he said plainly as though it were a fact he'd shared with her a thousand times, "I hope that's ok to say."

Not knowing how to respond, and feeling an instant fear welling inside of her, Anna just stared at Richard in shock for about half a minute and then finally managed, "I like you, Richard, I'm just not in the best place to be starting anything. I hope you understand. I thought I was, but I'm not. I'm confused, and I don't want to get hurt." Anna's brain had turned back on and it was in full force.

"I do understand, but that doesn't change how I feel about you, Anna. At the very least I'd like it if we could at least stay in touch. I think I did blow it the other day," Richard replied, his face showing the pangs of unrequited feelings.

"It's cliche, but it isn't about you. The truth is, I'm still healing and looking for answers. I'm not sure I'm going to find them on this mountain peak but I am hoping to learn a little more about myself on this trip. I'm not ready, Richard, I need to be in the right mindset to fully give myself to someone. I can tell that you are someone who deserves at least that much from someone. You're a remarkable man and I'm happy to have met you," Anna explained, hoping to give him some peace of mind knowing that it had nothing to do with him.

"If anyone gets that you're not ready, it's me. Trust me when I say that I understand you, Anna," Richard said as they stared, smiling at each other for a long time. Turning back to the towering mountains below them, they could see the rest of the group like ants making their way down the other side of the mountain. The teal lake meets with the monstrous glacier making its way between the mountains, optically flowing but actually retreating. Anna thought it a fine metaphor for her state of mind.

"I suppose we should continue the trek down, and try to catch up with the rest of the group," Anna said reluctantly, "I would rather stay here with you and gaze at this miracle. This place is absolutely captivating."

"Yes, I do suppose you're right. What does Ella think?" Richard left the decision-making to Ella, who was already bounding down the mountain like an avalanche.

"Alright then, I guess our mind has been made up for us. Let's go, but slowly, I'd rather spend a little more time looking at this view," Anna said, smiling, as they started back down the other side. The sides of the hazy lake came into view, nestled in between the mountains as they made their way lower down.

The wind, funneled through the pass, was picking up velocity and chilling them to the bone. Richard squeezed Anna's gloved hand on top of her hiking pole a few times and smiled at her, "Are you warm enough? It's getting chilly."

"Yes, I'm okay, Thank you. We're moving quickly now, so I'll warm up, and the wind probably won't be as bad further down at the base," she said optimistically, feeling the aching cold permeate her joints.

Half running down the mountain after Ella, Anna, and Richard were closing the gap and could see the last of the slower pack of hikers and the bright orange of Juan's cap. "Hey! We caught up with you finally," Richard yelled to Juan, catching his breath, "You guys are making a good pace, impressive."

"I thought you were right behind us, I'm so sorry!" Juan said apologetically.

"Well, we were, but then we stopped to take in the scenery and snap a few pictures. That peak offered some magnificent views, unlike anything I've seen in my life. It's like another planet," Anna said, shifting her pack and adjusting the straps again.

"Yes, I'm glad you're enjoying it. It's a strenuous climb, but I'm sure you'll agree that the reward is well worth the effort," Juan replied, looking back at them.

THE GROUP STAYED together for the next hour, trekking in unison at an even pace through the winding base trail, the looming dark landscape bearing down. There was one final section of the loop to complete before they would arrive back at the trailhead and finish the day's hike.

It was growing dark with clouds, and the cold dampness of evening was already settling in, but Anna felt warmth throughout her body hiking next to Richard. They had an unspoken secret, a tension she wondered if the others could see. Pangs of doubt started to tug at her stomach. Anna wanted more than anything to love Richard, but to open the floodgates to love would simultaneously open her up to a tsunami of pain if things were to go sour. She was too raw.

'What if this is just another repeat of Guatemala, and I am heartbroken and running for the hills in a few days? He lives a nine-hour plane ride away and he told me himself that he has no plans of settling down. What would that mean for us? Would this even have a future?' Anna's mind went on a rampaging tirade as she strode alongside Richard.

"Are you still warm enough, Anna? One more peak to go and then we're done for the day. I think we've got this in us!" Richard said encouragingly.

"Yes, I'm actually starting to feel hot with all of these layers," Anna smiled back.

"Ok, guys, we have just one more hour to go, and then we're back at the trailhead, and we can go back to town for a quick dinner before we turn in for the night. We're flying back to Buenos Aires tomorrow morning, so be ready in the lobby no later than seven," Juan yelled cheerily from the front of the pack.

FOR THE REMAINDER of the trek, Anna was lost in a whirlwind of spinning thoughts, her mind and heart waging war on each other like old foes. 'I'd felt only peace in that instance on the mountain, and now I'm lost in turmoil again. Am I making the right decision?' she thought in frustration. Anna wished that she could just be in the moment and not lament the future-projected consequences of living her romantic life with hopeful abandon. Outside, her mind was the beautiful serene mountain view; inside her mind was a raging forest fire.

It wouldn't be fair to attempt any form of relationship in the state that she was in. As much as she longed for companionship, Anna decided that she would remain friends with Richard. She could not complicate her life with a romantic situation. It would not be fair to Richard either; he deserved a focused person, and she was not. The purpose of the trip had been to heal, and although she had found her spirituality, and God, and faced her deepest fears, she was still in the process of healing her emotions and her mind.

THE GROUP MADE their way back to El Calafate after the hike, and it was their last night together before they would be leaving. Anna decided to join Sonja, Henrietta, and Ben at a restaurant called Bocal, where they would be treated to a live tango show. She hadn't spent much time with them but was hoping to get to know them a bit more before the end of the tour. She also needed some distance from Richard, to get perspective. Ben started ordering celebratory bottles of wine, and after her third glass, combined with the effects of the altitude, Anna was feeling slightly tipsy.

Anna's filter came off, and she started freely speaking about her half-baked theories of time travel, and religion, and eventually she slipped up and told them all about her reincarnated puppy, Ella, "Oh, yeah, and she is a puppy, but we've been together for the past seventeen years. She came back from beyond to be with me." She thought it was nice to have people to talk with about these things, and she couldn't understand why she had waited so long to open up about Ella. They seemed as supportive as Richard did.

"Very interesting, Anna. Tell us more," Ben had said, seemingly jokingly as he topped up her glass.

By the end of the night, Anna was feeling ill, "I'm not feeling very good. I wonder if I might have caught COVID. It's probably just a cold from being on the mountain, but I should go get a test to be sure. Can someone come with me?" she had asked.

"Ok, Anna, sure, let's go get you a covid test. I'm going to go call Juan to come with you," Sonja said as she got on the phone with Juan.

After half an hour, Anna saw Juan walking through the doors of the restaurant. "Oh dear, Anna, something must have happened to you when your kayak tipped over. We will take care of you, don't worry," he said.

"Yes, that's probably where I caught this cold, but I want a covid test to be sure. There are older people on the tour, and I don't want to put them at risk," Anna mumbled, feeling very embarrassed.

"We will need to go back to the hospital. It will be really quick, and we have an excellent healthcare system here. The pharmacies are closed, and you can only get the test at the hospital at this hour of the night. We'll be in and out in a matter of minutes, I'm sure, Anna," Juan said as they made their way down the winding streets of Calafate towards the hospital. Sonja and Henrietta had joined as well, on either side of Anna, holding her arms. Ella followed the group, intermittently growling in a low tone, and Anna couldn't tell what was wrong.

WHEN THEY ARRIVED at the hospital, Juan went away with a hospital worker in a blue shirt. Anna assumed she was going to get her COVID test after Juan. Through the small glass-pane window, she could see Juan chatting with the nurse for what seemed like a very long time before finally getting up and walking back to her, Sonja and Henrietta seated in the waiting area. "Ok, Anna, you're up. Bring your insurance information."

Anna wondered why they had spent so much time in the room together, but she pushed that thought aside as it was late, and she knew she had too much to drink. She wanted to get back to the hotel as quickly as possible to compose herself, drink water, and organize everything for the trip back to Buenos Aires. "Ok, Anna, you're up, come on," Juan called again, louder now as he was leaving the tiny room. Anna had a feeling that something wasn't quite right, but she couldn't place it. She followed another man named Felipe in a white dress shirt into a waiting room, and for some reason, Juan was accompanying them as well. Ella barked shrilly and tried to follow her as she started walking away, but Sonja had her leash gripped tightly.

"Hey, it's ok, Juan, I am not afraid of needles. I've already had my shot multiple times over the past two years; I've got this," Anna assured him with a smile and a tone of ease. 'I don't want someone with me for this,' she thought, 'This is making me more nervous having him here.'

"No, Anna, I want to be here for you. It's ok. Everything will be alright; don't worry. I'm right here by your side. Ella will stay with Sonja, and we really won't be very long, just a few minutes," Juan tried to reassure her.

FAITH   FEAR   FORTUNE

WHEN THEY SAT down in the waiting room, Juan and the attendant started speaking rapidly in Spanish, and Anna couldn't understand a word of it. They talked for maybe five minutes, and something in her gut told her to leave; something wasn't right with what was happening. 'What on earth could they be discussing for so long?' she thought with a pit in her stomach.

Another man came in wearing a brown sweater cloak over his shirt, and Anna could see a nameplate peeking out, 'Dr. Guido'. "OK, your name is Anna, right? It's ok, don't be nervous, Anna. Your friend here tells me that you can talk to animals, is that correct?" Dr. Guido asked her in thickly accented English.

"Yes, I mean, no. I'm just here to get my Covid test and vaccine, just in case there are some vulnerable people on the tour. Older people are hiking with us, and I don't want to make anyone sick. I'm also starting to feel terribly run down, and my throat hurts a lot. I didn't think anything of it until today, but I also woke up in a cold sweat last night," Anna explained.

"Oh, not a problem, Anna. Do you have family back home? How long have you been traveling?" Dr. Guido continued.

'I hate small talk with a passion,' Anna thought to herself, "I'm a journalist; I've been away for about three months, backpacking through Guatemala, Mexico, Costa Rica, and Columbia, and now I'm here taking in all of the beauty of Patagonia. I'm sorry, maybe I'm not making any sense but I'm a bit tipsy. Can I please just get a covid test and leave?" She humored him with as much small talk as possible but was getting very frustrated.

"Good! Very good. Ok, Anna. Juan here tells me that you're not feeling the greatest, so we're going to help you, don't worry!" Dr. Guido responded, "We will just monitor you until you're feeling better, just a few hours."

"I'd rather go back to the hotel with the group. I can keep myself separate; that's not an issue. It's probably just a cold, after all," Anna said, resisting his offer.

Juan was getting irritated, "Anna, can you please just stay for a while so we can all go back? We will meet you at the hotel tomorrow morning."

None of this was making any sense to Anna, but she was too tired to argue anymore, "Fine, I will stay no longer than two hours, and I want the test immediately," Anna finally agreed.

The three of them went into a room, and Dr. Guido went behind a curtain and started chatting with one of the nurses in Spanish. "That must be the nurse who is administering my test; let's get this over with so I can go back to the hotel and pack," Anna said to Juan.

"Of course, Anna, we will be here not a minute longer than needed," Juan agreed.

Dr. Guido and the nurse came into the room, and Anna started feeling strange. 'I know the protocol here is different from that of Canada's, but this is weird. Why all of this for a simple COVID test?'

she thought nervously.

"Here, Anna, just take these pills; they will help your throat and sniffles. Don't worry, it's ok. We have good medication in Argentina now," Dr. Guido said as the nurse stretched her hand toward Anna, producing two small white pills.

"I don't know what this is. Can you tell me what this is made of and what it's used for? What is the name of the medication?" Anna asked. She got the feeling they couldn't understand what she was saying, or maybe they just weren't listening to her.

"Anna, we really don't want to be here longer than absolutely necessary; we have an early morning. Can you please just cooperate so we can all get out of here? There isn't a rule book for this," Juan said in a stern manner that was seriously starting to give Anna the creeps.

The nurse, Juan, and Dr. Guido all stared at her. Anna felt backed into a corner. Deeply conflicted, she took the pills, swallowing hard, and felt immediate pangs of regret mixed with fear. 'Dear God, I hope I'm doing the right thing here. Please protect me,' she silently prayed.

"Do I get my covid test now?" Anna asked.

"Just relax, Anna, in just a few minutes. The nurse who administers the test is just finishing up with another patient. You Americans are always in a rush," Dr. Guido said.

'I'm Canadian,' Anna thought, but she wasn't feeling well and immediately knew something was very wrong. She closed her eyes and tried some deep breathing, thinking it might be the onset of a panic attack. Her heart was beating rapidly in her chest, and she was starting to sweat.

"You're ok, Anna! We are here to help you. What are you thinking right now, Anna?" Juan said in a soothing voice.

"Where is my Ella? Where is my girl?" Anna said as she passed out cold.

23

The Escape

Anna woke up in a hospital bed, alone although there was another bed in the room. The terribly scuffed walls were painted a sickly yellow, and cracks spread like creepy hands from the ceiling. The blue, sticky mattress had a white sheet half covering it, and the pillow was thick and hard. A bed rail had been pulled up on both sides of her, and Anna immediately felt like she was in some sort of a demented crib.

'Where am I? Where is Ella?' Anna thought as she slid down the bed and off the end of it. The porcelain tile was shockingly cold as Anna padded barefoot through the empty hallway. "Hello, anyone here?" she called softly.

"Hola, Anna, you ok?" a woman in a green uniform emerged from one of the rooms.

"Where am I?" Anna pleaded.

"You're in hospital. You had a situation, medical. You ok now, we take care of you. Not worry," the woman said in broken English.

"I don't need taking care of. Can you please tell me where my dog is? How did I get here? I came in for a covid test, and now I'm stuck in this place? Where are Juan and Sonja? They will be able to get me out of here," Anna was impatient and getting angrier by the moment.

"Oh no Anna, they are the ones who brought you here. They were concerned about you. You said funny things. Things that you're saying, they wanted you to be safe. My name is Euclid," the nurse said. Anna was having a difficult time understanding what she was saying.

"Euclid, nice to meet you. Maybe I had a few glasses of wine and said some nutty things, but no, I don't belong here. Someone could have even put something in my drink," Anna said emphatically. "I'm not angry at you; I'm sorry, I just don't know how people could do this to someone when they know they are away from family and friends alone in another country. They just left me here and got on a bus and left!" Anna was crying now as the nurse came over and hugged her.

"You will be out soon Anna, don't worry. It all going be ok," Euclid said reassuringly.

"I've been hearing that a lot lately, Euclid, and it doesn't seem to ever be ok," Anna said sobbing.

"Let's get you settled, and I'll show you around. The people here are kind hearts, Anna, you will be treated well," Euclid said as she took Anna's hand and started leading her down the corridor of the psychiatric ward.

'Someone from the tour is going to notice that I'm not around anymore. Someone will try to come and help me out. It's only a matter of time. I hope that Ella is alright; I'm sure she is missing me,' Anna thought hopefully. She was terrified for the first time in months. They were due to fly back to Buenos Aires today, and they had left her alone in an institution. She barely spoke the language.

BACK IN HER HOSPITAL room, Anna noticed that the other bed was made up with a pink blanket and there were personal items around it. 'I guess there is someone else here sharing with me,' she thought. She went over to the window. "Ella!" she yelled as she saw the beautiful ball of joy curled up under a hedge close to her room, "You beautiful girl, Thank you for staying close to me."

Ella sprinted over and started running back and forth, barking excitedly upon seeing Anna in the window. "It's ok, special girl, we're going to be together soon. Just a little while longer," Anna called as she looked over to see that the hospital had at least placed water bowls in the front of the building for the stray animals. 'Thank goodness she at least has some water to drink. I'll throw down whatever food I can get that might be good for her,' she thought.

This was a challenge that Anna was ready to solve; she wasn't about to let her fear control her again. She had to get out because Ella needed her, and there was no time to waste. They hadn't taken away her phone, which was a blessing, so she set to the task of contacting people she knew back home who could potentially help her devise a way out of the situation. The first person Anna would contact was her best friend Cole; she had, up until now, been sending Cole happy photos of her trip, and she wasn't sure how he'd react to her being locked up in a hospital.

Anna sent a message to Cole over text, "Hey Cole, so I've got myself in a bad situation. The tour operator had me locked up because I was saying some things that they thought meant I needed help. I messed up Cole and I don't know what to do. They have me in this psychiatric hospital in El Calafate."

HALF AN HOUR later Cole phoned Anna frantically, "I've heard some messed up shit in my life, but this takes the prize. Anna, are you serious? What the heck? Can they even do that to you?"

"Unfortunately, I am very serious. We were out for dinner, and I was running my mouth, and it rubbed someone the wrong way. I will admit I was saying some unconventional things, but taking me to the hospital and lying to get me here was way out of line," Anna explained, "I don't drink that much, and I know I shouldn't be drinking like I was, but it was just turning out to be such a fun evening and I was letting off steam. I should have just stayed in bed. I was feeling so sick after the hike. I really shouldn't have trusted them, Cole."

"Anna, what's done is done. We are going to find a way to get you out of this mess. Just try to stay calm for now," Cole reassured her.

"The worst part about all of this is that they took my bipolar medication away from me. I need that to remain stable. I don't know what kind of training these Argentinian doctors had, but they should at least know that you don't rapidly cut off someone's medication; it needs to be tapered. I've known my doctor for over ten years, and it took months to figure out what medication would be best for me. Now some kook in a coat has the power to undo all of that," Anna said frustratedly.

"Did you tell them about your diagnosis and your medication?" Cole asked.

"Yes, that was the first thing I said when I finally emerged from the sedation from whatever pills they had given me. I asked where my medication was and told them that I needed it. I'm already starting to feel dizzy and shaky. I'm really scared. I'm alone out here, I don't know the language, and I don't have anyone looking out for me or listening to me," Anna had broken out into tears now and was reaching through the phone across thousands of kilometers for support.

"We're gonna get you through this, Anna. Look at everything you've been through already, and you didn't quit. Just focus on the next step. Step by step, I promise you we are going to get you through this; you're not alone. I think the first thing we need to do is get ahold of the Canadian consulate in Argentina. They will be able to guide us on the next steps. You're distraught, so I'll take care of that," Cole offered.

"Thank you so much, Cole. I don't know how I would get through this without your support," Anna breathed a sigh of relief.

"We're also going to need to get a hold of your doctor and let her know what's going on and get her in touch with the doctors in El Calafate. They probably don't realize the dangers of stopping your medications so suddenly like this. Although they should! You're gonna get through this, Anna; you're not doing this by yourself. Like I said, every step of the way. I'm going to go jump on this and make a few phone calls," Cole finished, and they both hung up their phones.

Anna felt relief knowing that she wasn't going to be tackling this by herself, but she was still terrified of what would happen without her medication. She decided to go find the doctor to plead against taking away her meds. Asking a few nurses, one of them finally understood her and led her down a narrow hallway to the doctor's office and Anna stepped inside. "Hi Dr. Guido, my name is Anna and I am the woman who was admitted two days ago. I believe we spoke?"

Dr. Guido was a frail man, pale for the southern climate, and with a face covered in freckles. He was wearing another neutral-toned sweater cape, it seemed to Anna, to make him appear larger than life. His receding hairline made his pinched forehead wrinkles even more pronounced.

"Hi Anna, yes, I remember you. What can I do for you?" Dr. Guido asked.

"Well, I've had my medication taken away. It keeps my bipolar in check. Cipratox, and Lamaralit. I really need these medications as they help level out my moods. It's also very dangerous to stop taking these medications cold turkey, as they can cause serious, sometimes life-threatening side effects. I've been stable for seven years; it was seven years ago that I received my diagnosis," Anna reasoned with the doctor.

"Well Anna, you've got rashes growing on your skin and this is the starting point of Stevens-Johnson syndrome. We need to take you off your medications," Dr.Guido said arrogantly.

"Actually, it's eczema. I've had eczema my entire life, and it's triggered by stress. I'm very stressed out right now. I've suffered from it for as long as I can remember. I have had it for thirty-five years, long before I started taking the medication. I know for a fact that coming off of these medications cold turkey can cause terrible reactions. I know my body; I've tried coming off these medications before. I feel sick and I need my medication now," Anna continued.

"Actually no, Anna. You're going to need to come off of these meds and we'll slowly introduce them back into your system," Dr.Guido stated authoritatively, but from a place of little knowledge, Anna could now see.

Anna's gloves had come off, because if there was one thing she had learned since her diagnosis is that you have to advocate for yourself, "Absolutely not. I want my medication now. I'm not going through those withdrawal symptoms again. I'm taking nothing or I'm taking my medications. The liability for what happens to me will be on you and your institution." Anna raised her voice. Her face felt hot, her heart was racing, she felt dizzy and like punching a hole in the drywall of the putrid pink hallway walls as she dizzily walked out.

BACK IN HER room Anna sat on the white-railed hospital bed and let the tears flow. There was no way she was going to be bullied into submission by some halfwit doctor who was trying to throw his weight around, 'Where did he even train? Every doctor in the world should know that you can't take someone off of their medication.' After crying it out, Anna felt a lot better. She decided to walk around the ward of the hospital; she needed to find a way out.

The psychiatric ward was a perturbed island of inhabitants, and Anna wondered how anyone could get better there. Characters of all shapes were walking around like zombies through the pale halls, yelling gibberish, curled on beds and floors. The orderlies were easy to spot in their green uniforms, most of them looking sullen and tired. They didn't seem to take too much note of the patients. Anna needed to get out of this disturbing place before its greasiness stuck to her soul.

"Hi, you look American. Are you American? I'm Valentina," a young, pretty girl who looked very out of place, offered her hand to Anna.

"Hi Valentina, my name is Anna. That's my dog Ella outside the window there. I'm actually from Toronto, Canada. Have you heard of it?" Anna said.

FAITH   FEAR   FORTUNE

"Yes, I have heard of Canada; I want to go there. Everyone I talk to tells me that Canadians are such friendly people. Awe, beautiful puppy!" Valentina exclaimed as she smiled and waved out the window.

"Yeah, Ella is pretty awesome; we have been friends for a very long time. We've been on many adventures together, and I plan on going on a lot more when I can spring free from this place. I'm really worried about her being out there by herself," Anna said with a worried expression.

"This place isn't that bad. I'm used to it. I don't have a home, and my mother isn't nice to me," Valentina smiled, "You get to be my roommate." Anna thought that Valentina was probably the nicest person in the place and the only one who had approached her. She was grateful to have found someone who spoke her language and was kind to talk to.

Anna felt her phone buzzing in her pocket. "Richard! Oh, he must know that I'm here," Anna exclaimed out loud.

"Oh, is that your boyfriend or husband? Is he cute?" Valentina said.

"Yes, he is really cute, and well, no, he isn't my anything. We're just friends," Anna said reluctantly and with a pang of regret.

"I'm sensing something, hesitation is it? Don't let a good one pass you by, Anna, because you'll never know when a good one will come by again. No good ones for me yet, that's for sure," Valentina said, sighing.

"Well, I'm confused about Richard. I thought I wasn't ready, but am I ever happy when I hear from him, especially now. Do you think that sometimes everything in your life, good and bad, is eventually driving you to a specific end?" Anna mused out loud.

"I do believe that. It's what keeps me moving even on the dark days. I know that the sun will always come up the next day, and there will be a new opportunity to start again," Valentina had confidence in the future that Anna admired; she seemed so sure.

Anna read the very concerned message from Richard, "Anna! Juan told me that you're in the hospital. He wouldn't tell me what happened. Are you alright? What happened? Do you need anything? I've gone straight to Santiago, but I can try and help from here."

"I'm not sure what happened, Richard. I woke up in this place, and two days had passed. I have a feeling Juan is the reason I'm here, but it doesn't make sense to me. We were all drinking during the Tango show, I remember that, and I may have been tipsy. I'm not sure what I said that was so bad," Anna typed back.

"Oh, Anna, that's terrible. I will do whatever I can to help get you out. I'm a children's author, but my friend is a diplomatic aide, and I'm sure he knows someone at the embassy in Buenos Aires. I think he might be able to pull a few strings to help get you out quickly. Let me see what I can do," Richard replied instantly.

"Please, I would be so appreciative. Ella has no one, and I can see her pacing back and forth in front of the hospital looking for me. She has water but no love and no food. I think she is really scared for me, and it's breaking my heart," Anna typed back, happy to have someone else in her corner.

ANNA FELT TRAPPED and like the walls were caving in. 'How is anyone supposed to get better in this place? People screaming in the hallways, windows nailed shut, bread and water, plastic beds—it's like we are criminals. It's a really sad situation for people with mental health illnesses in El Calafate,' she thought. Anna had to get out of the hospital as soon as humanly possible. She decided that she couldn't wait for help; she needed to help herself.

She walked back over to the window. If there was a way that she could pry off the connector plates that were bolted across one of the windows, she would be able to open it and free herself. The only problem was that they had taken away most of the metal items that could be used, like tweezers and hair clips, but she did have a pair of nail clippers hidden in the pocket of her vanity bag.

Anna got her nail clippers, shut the door, and put her water bottle in front of the door. If someone started to open it, she would hear it tip over and be warned. She carefully edged the end of the clippers under the connector plate and started working it loose. To her delight, it didn't take long until the whole plate popped off. There were only five more plates to work off with the clippers.

Working as quickly as possible, she got all six of the plates off the window in just under ten minutes without being disturbed. She was hiding the disfigured plates under her mattress, just as she heard the water bottle fall and quickly stood up as Euclid, the nurse, walked in. "Hi, Anna, how are you doing? I just have to take your vitals and give you some pills," she said.

"I already have my prescription medication; they took it away from me. What are you giving me now? They don't know me or my condition; how can they possibly know what to prescribe?" Anna pleaded for her medication again.

"Only the doctor knows what this is. I'm sorry, Anna; I just following orders," Euclid replied.

Anna decided she would keep the medication under her tongue and spit it out when Euclid left. They were likely sedatives, the kind that had kept her in a lifeless state of passivity over the past two days. After Euclid left, Anna spat the medication into the toilet and flushed. She quickly reset the water bottle and went back over to the window.

It was going to take a Bondian effort to get her stuff out the window and shimmy down the gutter pipe, but she knew it could be done, and it needed to be done quickly to avoid being spotted. This was one of those times when traveling with a carry-on backpack was a blessing. They had taken a lot of her things, but she would just have to leave them behind. She would find a way to get her medication on the outside. She quickly penned a note to Valentina, "It was so great to meet you, but I need to get back to Ella as soon as I can. Thank you for showing me kindness when I was in darkness."

Anna pried the window open and turned around to make sure no one was there and that the water bottle that she had reset was still in its upright position. Pushing her backpack out the window first, she looked down to see it land safely in some shrubbery. Now she had to fight her fear and get down the three flights to the ground. Looking down from the ledge outside her window, Anna felt dizzy and had flashbacks of Calvin's suicide. Pushing these feelings down, she reached around and grabbed the gutter drainage spout with her left hand, secured her left foot on the ledge, and then pulled herself out and braced herself with four limbs against the wall. Slowly she started walking down the wall, keeping the tension with her arms. It was very challenging, but she was doing it.

ANNA PASSED the second floor and was making her way down the last flight when she passed a window with people on computers doing administrative work for the hospital. "¡Chica! Detenlo ahora mismo!" One of the older portly women yelled at her through the glass window. Anna didn't know what she was yelling but she knew there was no time to waste, letting go and dropping into the shrubs roughly.

The woman was opening the window as Anna looked up. She grabbed her backpack and raced across the road from the hospital down an alley. "Ella, let's go," she shrieked as Ella started racing after her.

To ensure she would throw them off track, she decided to go around in a loop and make her way in the opposite direction instead of going straight away from the hospital. Three blocks away, sweat was pouring down her face; she sprinted right and then right again, Ella following closely behind her. Seven blocks and five minutes later, she was on the other side of the hospital, and a park was in sight. Anna decided to hunker down for a few minutes in the dense bushes with Ella and regroup. She was entirely spent.

DIZZY AND CONFUSED, Anna was jolted by the familiar rush of the Existent Spirit and the gentle voice of Grandma Catharine, "Anna, get up now, it's time to get moving. You can't just lay around in these bushes all day; someone will find you. Take Ella and go to the airport. Follow the road to Santiago. Follow what was lost."

"What did I lose now? Never mind. Ok, Grandma. I'm going; I love you," Anna exclaimed, too dizzy and tired to think, adrenaline coursing through her veins.

If there was anyone she trusted right now, it was Grandma Catherine. Anna opened her phone and searched for a one-way ticket to Santiago, Chile. Within five minutes, she had booked her flight and summoned a taxi. "Let's move, Ella; we're going to Santiago!" Anna said as she grabbed her pack and jogged to the pickup location on the corner of Glacier Tosello and Glacier Agassiz streets.

FAITH   FEAR   FORTUNE

The taxi pulled up, and she jumped in, Ella pouncing up on her lap, making the scratches Anna had gotten from falling into the bushes burn. "Vamos al aeropuerto, gracias," she said to the man with graying hair in the front seat and kind eyes reflected in the rearview mirror.

AS THE TAXI made its way to the airport, she hugged Ella tightly and breathed a deep sigh of relief. It felt like she was finally free and in the clear. The only problem was that the doctors had taken away her medication so she would need to get it immediately when she was safely in Chile. Anna was already dizzy and lightheaded, and her hands were starting to tremble. At the airport, Anna rushed to the airline kiosk and checked herself in. She had five hours to wait for her flight, but that was a relatively short amount of time considering how last minute she had booked her flight. Once she was through airport security, she would be even safer. The security line was short, and she and Ella were through in just under fifteen minutes.

Anna decided to hide out in the corner of a small coffee shop and plan out her stay in Chile. 'I just need a place to stay,' she thought as she opened a home share app on her phone and found a beautiful studio apartment in downtown Santiago that even had a swimming pool. This would be the perfect place to regroup and recover from her ordeal at the hospital. After booking her apartment, she texted Cole to let him know that she was out of the hospital and alright, and then she texted Richard, "Hi Richard, I was able to get myself out of the hospital so please tell your friend that he doesn't need to contact the embassy and that I'm alright. Please thank him for me. I'm coming to Santiago today; it would be great if we could meet up next week if you're free, that is."

"Yes, of course, Anna. You're coming to Santiago! Really? How did you get out of the hospital? Did they let you go? Are you alright? Please, Anna, I'm really worried about you." Richard wrote back immediately.

"It's a long story, but no, they didn't let me out. I ended up leaving of my own accord, and I'm sure they're not too happy about it. Ella is with me, and we're at the airport already. I will fill you in on everything when I see you," she replied.

Richard seemed genuinely concerned about her, "Let's get together soon after you're settled in Santiago. I want to be there to support you; I can't imagine what you've endured over the past week, and I want to hear all about it, Anna. I'm here for you."

"OK, I will text you once I've checked into my place in Santiago and settled in. Thanks for the support, Richard; you're a great friend, and I want you to know how much I appreciate you," Anna finished the conversation.

THE AIRPORT was bustling with activity, and the volume of people milling about gave Anna and Ella the perfect cover. Over the next few hours, she had two coffees, a croissant, and a tuna sandwich, hiding out as long as possible in the back corner of the coffee shop. Twenty minutes before takeoff, she led Ella to gate twenty-two and sat near the attendant's desk. One of the men at the desk came over, "Miss, I see you have a service animal. You'll be able to preboard, and I can offer you the exit row so your companion has more room to stretch out. Would you like that?"

"Thank you so much! Yes, that would be amazing. This is Ella, and she is a great traveler. She won't make a peep, I promise," Anna said happily. Five minutes later, they were boarding the craft and settling into row twelve of the Boeing seven forty-seven. "We're almost home free, Ella; after takeoff, we'll be off Argentinian soil and on our way to Santiago and freedom," she said to Ella, who blinked back at her in quiet understanding.

People filed in along the corridor of the airplane, wedging their bags into the overhead compartments. The plane was packed to the brim, and Anna thought that it was by some miracle that she had managed to secure a seat for her and Ella. Someone was looking out for her; she thanked God and The Existent Spirit. A bouncy, chipper girl around seven years old, and her mother occupied the two seats beside Anna. The young girl was delighted to be next to Ella, "Can I please pet your doggie? What's her name?" the child piped.

"Her name is Ella, and you can absolutely pet her. She loves attention, and she is very sweet," Anna said, addressing the child and reassuring her mother. Ella put her paw up on the child's lap, and she laughed in delight, "Ha, my name is Henrietta; pretty miss Ella, you have the same colors as my favorite cookie." Henrietta nestled her head into Ella's neck and kissed her; in reply, Ella shut her eyes and snorted loudly.

"What cookie are you talking about, Henrietta?" Anna asked. Her mother responded, "Oh, she can't pronounce it yet, but she loves her Alfajores. They're those caramel and chocolate sandwiches."

"They sound delightful, Henrietta! Caramel is one of my favorite flavors in the entire world too. Ella also loves to lick caramel, but she can't eat chocolate. Dogs can't have chocolate," Anna responded.

"Why can't she have chocolate? It's so yummy," Henrietta asked curiously.

"Because it hurts their tummies and makes them very sick," Anna responded.

"Ok, no cookies for you then, fuzzy Ella," Henrietta said, giving Ella another delicate kiss on the top of her forehead.

As the plane taxied to the runway and lifted into the air, the weight slid off Anna's shoulders, and she finally felt free. The packed plane felt like safe padding against the terrible ordeal at the hospital and the end of a terrible chapter of her life. Over the two-and-a-half-hour flight, she rested her head against the plane window with Ella snuggled into her lap. The terrible Dr. Guido and the devious Juan couldn't get to her now; she could finally fall asleep.

FAITH   FEAR   FORTUNE

ANNA WAS JOLTED awake by the overhead lights coming on and the ding of the announcement system. "We have now started our descent into Santiago; please return your chairs to their upright and locked positions and stow any carry-on items under the seat in front of you." It was dusk, and the sun was going down on the horizon as the lights of the city, nestled in the mountains, were starting to come on. The plane glided gently into the airspace of the Santiago airport, descending carefully and touching down on the lit runway. It was a perfect landing, and Anna felt safe and secure. "Touchdown!" she said with a giant smile.

Lifting her bag from the overhead compartment, Anna and Ella exited the plane and were through Chilean customs in an hour. She summoned a taxi with her phone to take them to the intersection of Huérfanos and General Bulnes streets, and the condominium that she had reserved at the last minute. Santiago was a beautiful city; its fat palm trees lined the wide streets. Majestic churches with tall spires and crosses were everywhere. The area Anna had booked was densely populated and seemed safe. She arrived at the condominium and checked into apartment three-one-three, making her way through the courtyard beside the beautifully lit pool and lavender gardens.

Punching the eight-digit code into the lock pad on the door, it opened, and she entered, slumping against the back of the door and sobbing in relief as Ella licked the tears from her face. The apartment was small but beautifully appointed and had a comfortable queen-sized bed. Emotionally and physically exhausted, Anna fell onto the bed and passed out, fully clothed, Ella beside her.

24

The Manifestation

Once she woke and after taking Ella out for a bathroom break, Anna dumped the contents of her carry-on bag onto the floor and began organizing her clothing. It had all been shoved haphazardly into her small bag and was in a wrinkled, balled-up mess. She was going to spend two weeks in this place and needed to do laundry, find groceries, and get properly settled in. She noticed with delight that she had put a week of spare medication into a small travel pouch and was annoyed at herself for not having found it sooner. She promptly took her two medications and exhaled, exclaiming to Ella, "Thank God. Saved by overplanning."

Finding her pills was a huge weight off Anna's chest. Looking back on her journey, something in her heart made her believe that all of her trials and tribulations had conspired to bring her to Santiago. "Grandma Catharine had mentioned something about following what was lost, but hadn't I already found you in The Lost City, El? Was she talking about Richard? I had never had Richard in the first place, though," she said to Ella. The road had been long, and Anna had learned so much about herself. The life tests, the ones she willingly signed up for, and the ones that were thrown at her, she had stood up to them all.

As she put away her things, she searched for answers. "Where should I go next? What should we do?" Anna said out loud to a snoozing Ella on the bed. She had ended up here by chance and had no plans other than to meet up with Richard at some point. In the past, long walks always gave her time to think, so after putting away her things, she decided to get out for a walk with Ella and explore the city a bit. Her dizziness would fade once her medication kicked back in.

Opening her phone, she looked around for parks near her and saw that Santa Lucia Hill was only two kilometers away. The three-hundred-meter climb would give her ample time to clear her head, and she was looking forward to the fresh air and freedom after her ordeal at the hospital. She needed to get back to herself again. "Ok, let's go, El. This will shake the weight of the past week off of us," Anna opened the door and waited for Ella to leave first, locking the door behind them.

AS SHE CONTINUED to explore the city of Santiago on foot, Anna thought that it was beautiful. Most of the antiquated churches and historical buildings had been beautifully preserved, and there was a clear emphasis on parks and green spaces. As they strolled along the wide park promenade towards Santa Lucia Hill, the shade from the palm trees kept them cool and comfortable. It was a Tuesday morning, and the city was bustling with people rushing to work. When they arrived at the steps of the Hill, Anna motioned for Ella to go first and set the pace. Ella started running up the steps, and Anna ran after her, panting by the time they got to the fountain of Neptune. Pointing his trident towards the sky, seemingly directing them upwards, Neptune showed them the way.

Running hard, by the time she arrived at Terraza Caupolican, Anna was totally out of breath and barely keeping up with Ella. She figured it was remnants of her icy deep dive in Patagonia, another souvenir she would have rather left behind, along with her hospital bracelet. The last stretch was the highest point on the mountain. The stairs became treacherously slippery and jagged, reminding Anna of the twelve hundred steps at The Lost City, and she tried her best to keep up with Ella but started getting dizzy. Once again, the surroundings, and the distant miniaturized city below, faded into the background.

Anna heard the voice. "Anna, keep going. It's taken you everything within you to get to this point, and the road to Santiago has been long, but you are here now. What was lost will be found, and all will become clear when you reach the top of Santa Lucia, I promise you. It will all make sense soon, just take the next step, and the next step," Grandma Catherine said softly and encouragingly now, counting off each step.

Her head was buzzing, and she felt faint. How was it that she was feeling this way after only climbing two hundred or so steps? 'Just keep going, one foot in front of the other one,' she thought to herself. Anna felt that she was connecting with God, with the Existent Spirit and the essence of life as she took her last few steps of her climb up Santa Lucia. Finally, she reached the crest of the hill, and the entire city of Santiago was sprawled out and revealed to her. A panoramic vista, hazy snow-capped mountains in the distance, crisp structures surrounding the hill as the city was laid out.

She looked down at Ella, smiling up at her with her brown eyes and white teeth, and finally understood what it had all been for. There hadn't been any destination planned out for her; it had always been about the journey and about finding out what she was truly made of. It was a full-circle journey back to herself. For the first time in her entire life, Anna felt whole and present and completely at peace with where she was, and who she was. She was the Existent Spirit.

Life, it seemed, was inside her, connecting her to the outer world. She was a part of Ella, and Ella was a part of her. Most importantly, they were all a part of God. God was not separate; people were not separate; in an instant, Anna was awakened and conscious of the truth about the nature of life itself. There was nothing more than this moment, and the delight in that truth was all-encompassing.

Strangely, the image of Richard came to her mind, and it didn't frighten her; she felt only love and awareness of him. Anna wanted to be with him. Staring into the blue sky, the clouds, the mountains, and the richness and beauty of planet Earth were absorbed into her consciousness. Staring at the buildings and skyscrapers, the ingenuity of humankind filled her with pure delight and admiration. Life was a miracle. The Existent Spirit of God coursed through everything in front of her.

Anna's phone rang, and she picked it up. "Anna, you're never going to believe me, but I just got the urge to call you and see if you're alright. Are you alright, Anna? How are you doing?" Richard said.

"I've never been more ok in my life, Richard. There isn't anything in this world I'm lacking anymore. It's all there for all of us. I just feel so grateful for everything, the good, the bad, the pain, and the joy. It just all makes sense to me now," Anna said.

"Where are you, Anna? It's very windy, and I can barely hear you," Richard replied.

Speaking louder now, Anna replied, "I'm at the top of Santa Lucia Hill, and I was just thinking about you before the phone rang. Funny isn't it? I'd really love it if we could meet for dinner tonight if you don't have any plans, that is. I'd like to see you, Richard."

"You know, Anna, I was just thinking of inviting you for dinner, very strange indeed," Richard said happily, "I'd love to see you, Anna. I'm sorry if I scared you off back in Patagonia. I like you. I was confused, but I'm not anymore. I admire you, and I don't want to lose you from my life. Whatever shape that takes is fine with me."

"I'm through with overthinking my life and the people in it,

Richard. I want you in my life too. I understand that now. I'm not afraid of the consequences anymore. Even the worst consequences can't destroy me anymore, we are eternal, it doesn't matter. I'm living my life with love now," Anna said, laughing as though the greatest secrets of the universe had been revealed to her at that moment.

"Where are you? Can I come to see you tonight? Maybe five o'clock?" Richard replied.

"Yes, that's perfect. I'm at Huerfanos and General Bulnes. Call me when you're there, and I will come down and open the door. I will text you the address. I can't wait to see you, Richard," Anna said, ending the call.

There were no coincidences in life; she knew that now. She had manifested this in her heart, from a place of love, and because it had come from a place of love, it had come true. Anna did want to feel connected, and she did want love, and it was all coming to her now. That's what she had gained, that deep awareness and understanding that had previously been conditioned out of her. "Oh dear God, and life, and everything inside of me, and the Existent Spirit, thank you!" she cried, hugging Ella.

ANNA RACED DOWN Santa Lucia Hill beaming, Ella bounding happily behind her. By the time they got back to the apartment she was drenched in sweat and full of joy. Five o'clock couldn't come soon enough, but Anna tried to enjoy every moment of waiting. The hot water from the shower running down her face, the pleasure of it warming her body, the soft gentle music that was playing in the background, Ella snuggled into the lush pile bath mat. Everything all of a sudden seemed to be beautifully visceral and real. She was fully and completely present, and her mind wasn't ruling her emotions. For once, Anna's thoughts had been absorbed by pure delight in the present moment.

Anna blow-dried her hair straight, which was a rarity, and she put on some mascara and rose-tinted lip gloss, also a rarity these days. For the first time in a long time, she was genuinely excited and wanted to look her best. She chose a long flowing black skirt and a cream-colored top. Looking in the mirror, she loved the person who was staring back at her; she was fierce and beautiful and okay with feeling that way about herself. There was nothing arrogant about it; she finally loved the person she had become not for her appearance but for what she had accomplished and overcome. Anna felt completely in tune with who she was and what she wanted for the first time in her adult life.

Having some time to spare, she decided to make herself a cup of green tea and sit outside on the balcony. Smelling the tea, it was fresh and aromatic, and as she added the boiling water, she could see the leaf fragments mixed into the cup and swirl around. The air was a warm twenty-three degrees, and the gentle breeze caressed her body. Couples were swimming in the pool below her, happy. Anna breathed deeply and smelled the wafting scent of fresh lavender that had been planted around the apartment building, abundant in the garden, filling her lungs as she sipped her tea. She could hear children laughing and screaming, running up and down the street outside in delight. There wasn't a cloud in the sky; the day was surreally perfect.

RICHARD PULLED UP in a taxi earlier than five, equally excited about seeing Anna, and rang her on the phone, "Anna, I'm here, I'm downstairs."

"Ok, Richard, I'll be there in two minutes," Anna said, with butterflies in her stomach. Racing down the stairs with Ella closely behind her, she couldn't believe what she was allowing herself to feel, 'Is a man actually giving me butterflies? Is this what it feels like? All of the movies and novels I read talked about this feeling but I am finally experiencing it.' she thought. Racing downstairs to the lobby, Anna stopped and stared at Richard through the glass doors, taking in the moment, and smiling. He was wearing a pair of tan shorts, showing off his muscled runners' legs, and a white polo shirt. He looked like the best thing Anna had seen in years. Throwing open the door, Anna embraced Richard warmly and felt nothing but love.

"Richard, I wasn't ready before. I'm ready now. I'm sorry for what happened on the mountain, but I wasn't ready for you. I was too caught up in my thoughts. I didn't know who I was, but I understand who I am now, and it isn't what I thought about myself. I feel fully and completely what I want now, and that's to be next to you. You are a warm, wonderful soul," Anna said without any fear, and from a place of total love and awareness.

Staring into Anna's eyes, Richard looked so happy, his eyes reflecting her happiness. "Anna, I do love you. I love the woman you are. You're strong, brave, and resilient. I respect you, and I just want to care for you. I've never felt that in my life. I don't want anything more than to just be next to you. I can't explain it."

Ella had been standing behind Anna, watching it all unfold. Sensing Anna's happiness, she jumped up on Richard and licked his hand.

"I feel the exact same way about you. I love you too, Richard," Anna said, "But it isn't just you that I love. I love myself, and I love life again. There is so much hope in this place. I'm so glad I listened to the Existent Spirit and came to Santiago to find you. I can't wait for you to meet my family and friends back home, you're going to love them."

"I can't wait to meet them, Anna," he said warmly. I don't care what the future holds, or where we end up, as long as we have each other, we'll be just fine. Let's just live right now. The rest of it will fall into place because we do have forever, after all."

"Yes Richard, why don't we go upstairs and have tea? We have a lot of catching up to do."

To love the rush of uncertainty, stepping

into the unknown as my dim pulse

quickens. I journey far from

home to distant and alluring

lands, only to know myself

once more. Aching for the

vistas, and the feeling

of smallness in front

of the astounding

power of

nature.

"The Traveller" by Amy Shea

Acknowledgments

To my father, for giving me never-ending support throughout my childhood and into my adult life, making it clearly understood that failure was just an early stage of success. Thank you for always believing in me, and giving me the confidence to pursue my creative dreams, no matter how unconventional they seemed to be at the time. I am also so grateful to my family and friends for their support as I went through the challenging times I discussed in the novel. I would not have had the strength to get through this time in my life, or the courage to discuss these things as openly as I do if I did not have your unwavering support. Rose W, Claudio M, Mel G, Lisa N, Lola OP, Andrew H, and my brother Patrick Shea, thank you for being by my side physically or in spirit during the worst night of my life. I will never forget what you did for me during those times. Your impact cannot be overstated.

Last but not least, I would like to thank my favorite Irishman, Cahal Carmody for diligently editing this novel for me. Your steadfast friendship, attention to detail and care for my words will be forever appreciated. You have helped me realize a dream that at one time not so long ago, seemed too large to even imagine. Thank You to Cahal's dear wife Monica for supporting me in my journey, and making sure I had a soft place to land. The longest and most arduous journeys always begin with a single step, no matter how shaky that step may be. Go raibh maith agat, muchos gracias, and thank you sincerely.

Thank you to the many furry friends I've met on my travels around the sun

Caramel the dog, The Lost City, Columbia

Bella, also known as 'Ella' the dog, Toronto, Canada

Author Biography

Amy C. Shea is a native of Toronto, Canada who has been an avid hiker since the age of five. A self-proclaimed animal lover and naturalist, she enjoys adventure travel, being in nature, and sailing on Lake Ontario. Amy's background is in design, management, and user experience, and she has worked for some of the largest brands in the world, including Sony, Unilever, and Shopify. A passionate writer, this is her first published novel, inspired by the hardships and challenges she faced in her own life.

With this novel, Amy's goal is to bring comfort to those who may be victims of domestic abuse, and also to grieving pet owners like her. She hopes to spark the imagination and hope of her readers, entertaining them, and opening them up to the possibilities of life beyond just our limited physical form. The book has a large spiritual component, which is reflective of her own beliefs that were shaped through experiences in her life.

You can connect with Amy: **instagram/faithfearfortune**

Manufactured by Amazon.ca
Acheson, AB